PRAISE FOR *IT AIN'T NECESSARILY SO*

"Lewontin is the Voltaire of the Age of the Absolute Gene. With clarity, wit, and a practiced eye for high-tech humbug, he exposes the fallacies and impostures of all sorts of biological determinism, from evolutionary psychology, the Bell Curve, and anthropometrical brain weighing to gene therapy, fixed-function brain mapping, DNA vitalism, gendered intelligence, and hereditarian twin study. As biology becomes our dominant science and Darwinism our master narrative, this is a necessary book."

— Clifford Geertz

"Lucid and based on deep understanding in many domains, these essays are not only highly informative but also invariably stimulating and provocative, their value enhanced by new material of the same impressive quality and significance."

— Noam Chomsky

"What is unusual about Lewontin and his eloquent critique is that, apart from being extremely subtle and intelligent, he is a working biologist and a wonderfully stylish writer.... If you read only one book on genetics this year, make sure it is this one."

— *The Sunday Times* (London)

"A very fine and important book, and a very necessary corrective to all sorts of popular fallacies."

— *The Guardian*

"His writing is consistently elegant and readable, frequently funny, and abounding with provocative remarks."

— *Nature*

"Well-written, insightful, and a useful reminder of the complex issues still unresolved in the biological sciences."

— *Kirkus*

It Ain't Necessarily So: The Dream of the Human Genome and Other Illusions

It Ain't Necessarily So:
The Dream of the
Human Genome and
Other Illusions

by

Richard Lewontin

SECOND EDITION

NEW YORK REVIEW BOOKS

New York

IT AIN'T NECESSARILY SO:
THE DREAM OF THE HUMAN GENOME
AND OTHER ILLUSIONS

by Richard Lewontin

This edition published in 2001
in the United States of America by
The New York Review of Books
1755 Broadway
New York, NY 10019
www.nybooks.com

Library of Congress Cataloging-in-Publication Data

Lewontin, Richard C., 1929–
   It ain't necessarily so: the dream of the human genome and other
illusions / by Richard Lewontin.
      p.   cm.
   Includes bibliographical references.
   ISBN 0-940322-10-2 (acid-free paper)
   ISBN 0-940322-95-1 (pbk.: acid-free paper)
   1. Nature and nurture. 2. Human genetics. 3. Genetics. 4. Human
biology. 5. Biology. I. Title.
QH438.5 .L49 2000
576.5—dc 21   99-056280

ISBN 0-940322-95-1

Printed in the United States of America on acid-free paper.

October 2001

To Mary Jane who has taught me so much that is really so

Quantitative Zoology (with George Simpson and Anne Roe)
(1960)

The Genetic Basis of Evolutionary Change (1970)

Human Diversity (1982)

Not in Our Genes (with Steven P. R. Rose and Leon J. Kamin)
(1984)

The Dialectical Biologist (with Richard Levins)
(1985)

Education and Class (with Michael Schiff)
(1986)

Biology as Ideology (1991)

Inside and Outside (1994)

Gene, Organismo e Ambiante (1998)

The Triple Helix (2000)

Acknowledgments

The greatest debt an essayist has is to his or her editor, and I have acknowledged how much I owe to Robert Silvers elsewhere in this book. I simply reiterate here my admiration for his editorial perspicacity. The reader of a book may not realize, however, that its existence owes an immense amount to the essential work of people who go over the manuscript and its various preliminary printed versions in exquisite detail and who bring the book into the consciousness of a reading public. My considerable gratitude must be expressed to Borden Elniff, Karla Eoff, Michael Shae, and Catherine Tice for turning what was written into a publishable book and to Rea Hederman for actually publishing it.

Finally, the articles on which the book is based could not have been produced in the first place without the organizational skills and interventions over many years of Rachel Nasca, to whom I am, as always, extremely grateful.

Contents

I'm preachin' dis sermon to show
It ain't . . . necessarily so!

INTRODUCTION

CUT IN STONE over the main entrance to Emerson Hall, the seat of Harvard's Philosophy Department (and, in former days, of its Psychology Department as well), is the Psalmist's question,

WHAT IS MAN THAT THOU ART MINDFUL OF HIM?

What has changed since Emerson Hall was built in 1905 is not the question, but the identity of "THOU," which is left tantalizingly ambiguous by the stonecutter's block capitals. Changed, too, is the departmental responsibility for the inquiry. It is now the Department of Molecular and Cellular Biology (located with characteristic Harvard prudence on Divinity Avenue) that investigates the problem of the nature of human beings with automatic DNA sequencers and microchips, rather than by faith or philosophical inquiry. Indeed, the composer of Psalm VIII already foresaw the relevance of genetics to his question when he added, as his next line,

And the son of man that Thou thinkest of him?

A great deal can be learned by reading books. Much of what can be learned from the content of a book is, of course, untrue, but there is a reliable source of information on the front and back of the title page. Books are commodities meant to be bought, which means that they are meant either to be read or displayed on coffee tables by a sufficient number of purchasers, or at the very least to be held aloft on the television screens of a large number of viewers. The title and subtitle of a book of nonfiction, together with the year of its publication, is then immensely revealing of the state of public consciousness at a particular time.

The essays collected here were originally book reviews written over a period of seventeen years for what rapidly became, after its first appearance in 1963, the leading publication for a wide intellectual public. While intended as a journal of book reviews, *The New York Review* has served a much more general intellectual purpose. As the editors so well understand, book reviews can be a marvelous vehicle for engaging readers in a serious consideration of a variety of questions of interest to them. In republishing these essays a problem of rhetoric arises. Book reviews are, to some extent, a form of journalism. Necessarily, especially in writing about science, an essay reflects the state of knowledge and of the world at the time the essay was composed. But there are new developments in science and new events that give an old essay a certain dated quality, requiring some *aggiornamento*. At the suggestion of Robert Silvers, who has been the editor of these reviews over the years, I have left them in their original state and have added brief postscripts where necessary to bring them up to the present. Thus, the reader has the advantage of some historical perspective. Even when a very short time has elapsed since its first composition, the matters discussed in a review may have an eventful history. That is certainly the case for the matter of cloning, which has developed considerably even since October 1997.

These essays are all about various aspects of biology, especially human biology. The fact that *The New York Review* has published so many essays on biology, not only those of mine contained in this book but of many other scientists, such as S. J. Gould, Peter Medawar, Steve Jones, and Max Perutz, reveals the importance of biological questions in the consciousness of a reading public and of the dominant role that biology has come to play in the collection of activities called "science." But it also reveals the perspicacity of Robert Silvers. First, it must be understood that it is he who suggested nearly all the books for review and who realized the possibilities latent in them. Second, his taste as an editor is exceptional. Suffering from a certain authorial arrogance, I usually resist the intervention of editors. I make an exception for Silvers. When a query has been penciled in the margins of a galley proof, it has always been right on. He has been unerring in detecting opacities and for suggestions of how further detail would make the rough places plain. The only unresolved disagreement we have had is over his unreasonable refusal to allow the use of the word "problematic" when referring to the problematic of an intellectual discipline.

It is obvious to all that, as the power to manipulate the physical world has increased, the locus of inquiry into the nature of the world has passed from the realm of philosophical argument to the domain of natural science, a passage that has been accelerating since the seventeenth century. Our ordinary experience of the physical world has been made for us by physicists. Even for those phenomena that are beyond our power to manipulate, we expect, at least, some advance notice of catastrophes from the meteorologists, volcanologists, seismologists, and comet watchers. What is less obvious, because it is a more recent historical phenomenon, is the way in which biological science has displaced the classical physical sciences, both in prestige and economic power in the community of science and in the public consciousness as well. The final enthronement of physics as Science Triumphant was proclaimed on August 6, 1945, with a blast heard around the world, immediately convincing the smartest high school science students that they wanted to become nuclear physicists. The importance of the physical sciences, and in particular their practical embodiment in engineering, was then further emphasized by the appearance of Sputnik in 1957. Even as late as 1960, when my children were in school in Australia, biology was a girls' subject, not considered a fit study for clever boys. Not only natural science itself but concerns with the history and especially the philosophy of science were dominated by problems in the physical sciences. In the year following the appearance of Sputnik there were only two short articles about biology in *Isis*, the official organ of the Society for the History of Science, and none in the journal *Philosophy of Science*. But we have changed all that.

Beginning slowly in the 1950s, at the peak of the prestige and success of the physical sciences, physicists and chemists began to migrate into biology, becoming the founders of modern molecular biology. This seemingly paradoxical movement was in part a reflection of the hubris of physicists who, dizzy with success at blowing things up, were in no doubt that the kind of science used to split the atom could solve the much more complex problem of dissecting protoplasm. But it came also from a growing feeling that the very success of the physical sciences meant that all the really worthwhile problems that could be solved had already been solved, and that the only interesting field left for a scientist was biology. Beginning at the same time, and accelerating immensely after Sputnik, state expenditures for basic science increased exponentially, making available previously unimagined sums for biological research.[1] Biology was not only interesting and important, but one could make quite a good career out of it. The increasing dominance of biology within science over the last forty years has also produced a change in the interests of historians, philosophers, and

---

1. For a detailed history of the immense growth of state expenditure on science and its role in the growth of academic science, see R. C. Lewontin, "The Cold War and the Transformation of the Academy," in *The Cold War and the University*, edited by André Schiffrin (New Press, 1997).

sociologists of science. Not only are their standard journals filled with articles about biology, but we now need specialty publications like the *Journal of the History of Biology* and *Biology and Philosophy*, just to keep up with the demand for space.

The change from science as physics to science as biology is not merely a redirection of academic lives. It reflects our general view of what we want to know about the world. We may be interested, in a detached way, in how long ago the Big Bang resounded, or in how many kinds of indissoluble little bits make up all matter, but what we really want to know is why some people are rich and some poor, some sick and some well, why a woman can't be more like a man, and why I can't live to be a sexually active centenarian.[2] In consciousness, as in science, the animate has come to dominate the inanimate and, in particular, it is now widely believed that the main question of scientific investigation ought to be not what constitutes matter but what it is to be human. Nothing illustrates this change in priorities better than Congress's cancellation of the immensely expensive supercollider project aimed at finding the ultimate building blocks of all matter, while approving the immensely expensive Human Genome Project aimed at describing the complete sequence of DNA that is said to build a human being.

---

2. It is hard to resist quoting Steven Weinberg here: "We don't study elementary particles because they are intrinsically interesting, like people. They are not —if you have seen one electron, you have seen them all." "The Revolution That Didn't Happen," *The New York Review*, October 8, 1998, pp. 48–52.

The books reviewed in these essays turn out, unintentionally, to cover the development of modern biology from Darwin to Dolly. In the long sweep of the history of modern science, biology developed slowly. Scientific, that is, mechanistic experimental, biology began in the seventeenth century with Harvey's description of the circulation of the blood and Descartes's *bête machine*, but not a great deal more happened for another two hundred years. Even the person we now regard as the preeminent biologist of the nineteenth century, Charles Darwin, owed his instant fame to a speculative theoretical essay on natural history, while, in contrast, Mendel's carefully designed quantitative experiments on heredity went unnoticed until 1900.[3] We need to remind ourselves that the possibility of spontaneous generation of living forms from inanimate matter was still an open question until it was agreed that Pasteur settled it in 1860. Moreover, despite his great contributions to public health and wine making, Pasteur thought of himself as a chemist (he started as Professor of Physics at Dijon and then became Professor of Chemistry in Strasbourg), and his most important contribution to basic science was his founding of the science of stereochemistry, the study of the alternative three-dimensional shapes of molecules. In the middle and later part of the nineteenth century, in the service of medicine, a great deal was learned about the physiology of microorganisms and vertebrates, but the only large project of biological research

---

3. See Chapter 2 and Chapter 3.

that was self-consciously designed to bring the living world into congruence with the mechanistic principles developed for the inanimate world was in the study of embryology. The intellectual program of the German school of *Entwicklungsmechanik* was to provide an entirely mechanical explanation for the mysterious and seemingly goal-directed process of the development of a highly differentiated adult organism from a single cell.[4]

This program, as yet unfulfilled, continues to give form to a large part of modern biology. It is the raison d'être of the Human Genome Project.[5] For nineteenth-century biology the program of providing a mechanical explanation of development was more than a question of producing a coherent explanatory scheme for living organisms. It held out the promise of success in the ultimate goal of biology, to produce living organisms artificially in the laboratory. Indeed, for Jacques Loeb the production of life in the laboratory was, by definition, what it means to understand it.[6] Loeb never got further than inducing an egg to develop without fertilization, but cloning was the next step.[7]

The ambitious program of the nineteenth century, to make biological phenomena simply the extension of the physical world, has been a remarkable success in the twentieth. The physiology and metabolism of

4. See Chapter 4.

5. See Chapter 5.

6. See Chapter 4.

7. See Chapter 8 and a subsequent exchange.

organisms and of the cells that make them up is well understood at the level of the shape and chemical behavior of molecules. Not only the statistical regularities of inheritance, but the cellular and molecular dynamics that underlie them, are known. We have very likely scenarios for the origin of living forms from the primordial slime, and the subsequent processes of organic evolution can be understood as the outcome of simple processes like mutation, natural selection, and random births and deaths.

Two major domains remain to be satisfactorily included within the mechanist program. One, ironically, is the very problem that nineteenth-century biology took to be the major challenge for a mechanistic science of life, the problem of the development of form. A great deal is known about the genes that encode various chemical signals in development and about how the network of signals is hooked up, but we do not have the faintest idea about how all of this is turned into the shape of my nose. We do not even know how to ask the question in a useful way, although some interesting models have been made of how chemical signals might affect the arrangements of cells.[8] The other field of immense ignorance and conceptual poverty is the problem of understanding the central nervous system.[9] What is the mapping between the physical states and connections of brain cells and mental states? It is not even clear that the same mental

8. See the second part of Chapter 4.

9. See the last part of Chapter 3.

states are mapped into the same brain locations in different individuals or even in the same individual at different times. There are some crude localizations to big chunks of the brain of general categories of perceptions and mental states, but nothing at all can be said of the physical processes that led me to write the previous sentence as opposed to the infinity of sentences I might have written. Indeed, we do not know why a monkey, given a keyboard, could not have done the same thing. Yet no biologist is in any doubt that development and central nervous function are consequences of mechanical and chemical forces among structured assemblages of molecules and could, in principle, be so described. We feel no need of mysterious new fundamental forces characteristic only of living organisms, of an entelechy that drives the developing embryo to its destined final form, or of a ghost in the machine of the brain. No biologist now doubts that organisms are chemico-electrico-mechanical systems. Our problem is that, in contrast to other domains of the physical world where a few strong forces dominate phenomena, the organism is the nexus of a very large number of weakly determining causal pathways, making it extremely difficult to provide complete explanations.

The success of the program of physicalizing biology has encouraged the program, also inherited from the nineteenth century, of biologizing the psychic and the social. After all, if thoughts, attitudes, temperament, and culture are manifestations of the activity of a physical organ, the brain, does it not then follow that the *causes* of thoughts, attitudes, temperament, and culture are identical with the causes that specified that physical organ? More specifically, it is easy to think that if organisms are largely the consequences of the genes that they have inherited, then the similarities and differences of organisms are the consequences of similarities and differences in their genes. Anthropology, sociology, psychology, political science, economics, linguistics, moral philosophy become branches of applied biology, in particular genetics and evolution. It was widely assumed in the nineteenth century that human character was in the blood, and as vague notions of "blood" gave way to more articulated theories of inheritance, theories of the inheritance of character became more seemingly scientific.

In his preface to the Rougon-Macquart novels, written in the thirty years before the rediscovery of Mendel's original work, Emile Zola assured his readers that "heredity has its laws, just like gravitation."[10] Those laws turned out to be very different from what Zola's contemporaries believed them to be, but the force of Zola's claim for the heredity of character has remained powerful in explanations of human events. Now that we know the true laws of heredity, it must be that we know the real laws of the formation of the human psyche. What has changed since the nineteenth century is the substitution of genes for blood and the fusion of modern genetics with the Darwinian theory of evolution by natural selection. If Europeans have dominated the darker races, it is because evolution in colder climes

10. See my essay on the Rougon-Macquart novels, "In the Blood," *The New York Review*, May 23, 1996, pp. 31–32.

has led to a genetic superiority in mental and moral traits.[11] If men dominate women and are unfaithful to them, it is because the evolution of the human species has favored those genes that make men more aggressive and more sexually promiscuous, while producing gentle, nurturing, constant females.[12] If I send $100 to the Greater Boston Food Bank and thus forgo a dinner at Chez Robert, it is because the effect of a gene for altruistic behavior toward relatives is a bit imprecise and leads me to a less discriminating self-sacrifice.[13] It is all a matter of the proper genetic strategies for leaving more offspring. The strain of simplistic scientism that characterized social theory from the beginning of the nineteenth century continues to infect it today. The extraordinary increase in the knowledge of the mechanistic details of inheritance and evolution has only changed the language in which that scientism is expressed.

The successes of the natural sciences in explaining the physical and biological world have affected not only the content of explanations of social phenomena but the image of how we are to go about investigating them. Studies of human society become "social sciences" with an apparatus of investigation and statistical analysis that pretends that the process of investigation is not itself a social process. The problem for the investigation of society is that often, although not always, the evidence we need resides only in people's heads and

---

11. See Chapter 1.

12. See Chapter 6.

13. See Chapter 9.

the only way to get the information is to ask them. But what if they do not tell us the truth?[14]

It is not only in the investigation of human society that the truth is sometimes unavailable. Natural scientists, in their overweening pride, have come to believe that everything about the material world is knowable and that eventually everything we want to know will be known. But that is not true. For some things there is simply not world enough and time. It may be, given the necessary constraints on time and resources available to the natural sciences, that we will never have more than a rudimentary understanding of the central nervous system. For other things, especially in biology where so many of the multitude of forces operating are individually so weak, no conceivable technique of observation can measure them. In evolutionary biology, for example, there is no possibility of measuring the selective forces operating on most genes because those forces are so weak, yet the eventual evolution of the organisms is governed by them. Worse, there is no way to confirm or reject stories about the selective forces that operated in the past to bring traits to their present state, no matter how strong those forces were. Over and over, in the essays reproduced here, I have tried to give an impression of the limitations on the possibility of our knowledge. Science is a social activity carried out by a remarkable, but by no means omnipotent, species. Even the Olympians were limited in their powers.

RICHARD LEWONTIN

---

14. See Chapter 7.

Chapter 1

THE INFERIORITY COMPLEX

"The Inferiority Complex" was first published in
The New York Review of Books *of October 22,
1981, as a review of* The Mismeasure of Man, *by
Stephen Jay Gould (Norton, 1981).*

THE FIRST MEETING of Oliver Twist and young Jack Dawkins, the Artful Dodger, on the road to London was a confrontation between two stereotypes of nineteenth-century literature. The Dodger was a "snub-nosed, flat-browed, common-faced boy... with rather bow legs and little sharp ugly eyes." Nor was he much on English grammar and pronunciation. "I've got to be in London tonight," he tells Oliver, "and I know a 'spectable old genelman lives there, wot'll give you lodgings for nothink...." He was just what we might have expected of a ten-year-old streetwise orphan with no education and no loving family, brought up among the dregs of the Victorian *Lumpenproletariat*.

Oliver's speech, manner, and posture were very different. "'I am very hungry and tired,'" he says, "the tears standing in his eyes as he spoke. 'I have walked a long way. I have been walking these seven days.'" Although he was a "pale, thin child," there was a "good sturdy spirit in Oliver's breast." Yet Oliver was born and raised in that most degrading of nineteenth-century institutions, the parish workhouse, deprived of

4

all love and education. During the first nine years of his life he, "together with twenty or thirty other juvenile offenders against the poor-laws, rolled about the floor all day, without the inconvenience of too much food or clothing."

Where amid the oakum pickings did Oliver find the moral sensitivity and knowledge of the English subjunctive that accorded so well with his delicate form? The solution of this, the central mystery of the novel, is that Oliver's blood was upper-middle-class, though his nourishment was gruel. Oliver's whole being is an affirmation of the power of nature over nurture. It is a nineteenth-century prefiguration of the adoption study of modern psychologists, showing that children's temperaments and cognitive powers resemble those of their biological parents whatever may be their upbringing. Blood will tell.

Dickens's explanation of the contrast between Oliver and the Artful Dodger is a form of a general ideology that has dominated European and American social thought for the last 200 years, and is the central concern of Stephen Jay Gould's book—the ideology of biological determinism. According to this view, the patent differences between individuals, sexes, ethnic groups, and races in status, wealth, and power are based on innate biological differences in temperament and ability which are passed from parent to offspring at conception. There have, of course, been countercurrents of "environmentalism" emphasizing the malleability of individual development and the historical contingency of group differences, but, with the exception of Skinnerian behaviorism, all modern theories of social development have assumed an irreducible nontrivial variation in innate abilities among individuals and between groups. Occasionally, the political consequences of extreme biologism have been so repugnant that environmental and social explanations of group differences have held temporary sway. So, the practical application of biological race theory by the National Socialist state discredited biological theories of racial and ethnic superiority for about thirty years, but by 1969, with the publication of Arthur Jensen's monograph *How Much Can We Boost IQ and Scholastic Achievement?*, it was once again not only respectable, but even popular, to argue that blacks owed their inferior social position to their inferior genes.

Because biological determinism is a structure of social explanation that uses basic concepts in anatomy, evolutionary theory, genetics, and neurobiology, often in a corrupted form, its critique demands the powers of a historian of ideas and a professional biologist. Because the scientific methods and concepts involved are rather abstruse, criticism also requires a first-class writer. Fortunately, Gould is a professional historian, an evolutionary biologist and anatomist of great accomplishment, and a master at explaining science. *The Mismeasure of Man* is his examination and debunking of the scientific face of the fiction of Oliver.

Dickens's view of the origin of human variation was hardly exceptional; it permeated nineteenth-century literature: at times it appeared only incidentally as part of the substrate of unspoken assumption as, for example, in *Felix Holt*, when Esther Lyon is set to learning French on the assumption that her French ancestry will make it easy for her. At others, it is a central preoccupation, as in Eliot's *Daniel Deronda*. Daniel, the adopted son of a baronet, is a typical young English milord, whom we first meet at a fashionable Continental gambling spa. But then, mysteriously, in his young manhood, he develops an interest in things Hebrew, falls in love with a Jewish girl, becomes converted. The reader is not entirely astonished to learn that Daniel's mother was, in fact, a Jewish actress. The Law of Return, it seems, is only an expression of the inevitable.

A preoccupation with the power of blood was not simply what the French know as "the madness of the Anglo-Saxons." Eugène Sue, the most popular French author of the mid-nineteenth century, created in *Les Mystères de Paris* the archetype of the noble prostitute, somehow unsullied and saintly in the midst of her sordid existence. She was, of course, the abandoned child of a morganatic marriage. Among the *goyim* at least, true character apparently can be transmitted through the paternal line. But it is in the Rougon-Macquart novels of Zola that biological theories of character are given their most careful articulation. The Rougons and Macquarts were, it will be recalled, the two halves of a family descended from a woman whose first, lawful, mate was the solid peasant Rougon, while her second, illicit, lover was the violent, unstable Macquart. From these two unions arose an excitable, ambitious, successful line, and the depraved, alcoholic, criminal branch that included Gervaise and Nana. When Coupeau, Gervaise's husband, is admitted to the hospital for alcoholism, the examining physician asks him first, "Did your father drink?" As Zola says in his preface to the cycle, "Heredity has its laws, just like gravitation."[1]

Zola's "experimental novels," as he called them, were the outcome of developments in physical anthropology as a scientific, materialist discipline, developments to which the first part of *The Mismeasure of Man* is devoted. In America, Samuel Rogers Morton had, in the 1830s and 1840s, measured large numbers of skulls of different human groups, including long-dead Incas and ancient Egyptians. The Anthropological Society of Paris had been founded in 1859 by Paul Broca, the leading European exponent of the theory that high intelligence and character were a consequence of larger brains, so that the mental qualities of individuals and races could be judged from the sizes of their skulls. The appearance, in the same year, of *On the Origin of Species* gave rise to an evolutionary view of human differences that placed each physical type on an ascending scale of progress from our apelike ancestors. In particular, criminals were seen as atavisms, apelike in both mind and body, but in a variety of forms, so that the founder of criminal anthropology, the Italian Cesare Lombroso, could tell a murderer from an embezzler at a glance. But Broca and Lombroso were only

---

1. Emile Zola, preface to *La Fortune des Rougons* (Verboeckhoven: Librairie International A. Lacrois, 1871).

the inheritors of a long tradition that began with the natural philosophers of the eighteenth century.

The reductionist materialism of Descartes's *bête machine* and La Mettrie's *homme machine* led inevitably to the anthropometry of Broca and Lombroso. If mind is the consequence of brain, then are not great minds the products of great brains? Indeed, phrenology was a perfectly sensible materialist theory. Since acquisitiveness is a product of a material organ, the brain, then highly developed acquisitiveness should be the manifestation of the enlargement of one region of the brain. On the not unreasonable (although factually incorrect) assumption that the skull will bulge a bit to accommodate a bulge in the cerebral hemisphere, we might well expect an enlarged "bump of acquisitiveness" among the more successful members of the Exchange, not to mention Jews in general.

Moreover, less developed races should have less developed brains, women should have smaller cranial capacities than men, the lower classes more sloping foreheads than the bourgeoisie. Thus one should be able, by the appropriate physical measurements, to characterize the mental, moral, and social attributes of individuals and groups. There are, however, two problems with this theory. First, there is the factual error. Despite all claims to the contrary, there are no differences in brain size or shape between classes, sexes, or races that are not the simple consequence of different body size, nor is there any correlation at all between brain size and intellectual accomplishment. Second, there is the

conceptual error. Intelligence, acquisitiveness, moral
rectitude are not *things*, nor the natural attributes of
things, but mental constructs, historically and culturally contingent. The attempt to find their physical site in
the brain and to measure them is like an attempt to
map Valhalla. It is pure reification, the conversion of
abstract ideas into things and their natural properties.
While there may be genes for the shape of our heads,
there cannot be any for the shape of our ideas. It is with
an exposure of these two errors of biological determinism that Gould's *The Mismeasure of Man* is largely
concerned.

The first problem is to explain how the zoologists and
anthropologists of the nineteenth century could find, so
consistently, that, for example, the brains of whites are
significantly larger than the brains of blacks when, in
fact, there is no difference between them. The answer
seems to be, according to Gould, that the most eminent
zoologists and anthropologists simply rigged the data.
When Samuel Morton, in his *Crania Americana* of
1839, showed conclusively that American Indians had
smaller craniums than Caucasians, he did so by including a large number of small-brained (because small-
bodied) Inca skulls in his Indian sample, but at the
same time excluding a number of Hindu small-skulled
specimens from his Caucasian sample. When Gould
recalculated the data using all of Morton's measurements, the difference between Indians and Caucasians
disappeared. Paul Broca, faced with some very small
brains of some very eminent professors, invented ad

hoc corrections for age and postulated disease. As a last resort he appealed to the imperfection of institutions:

> It is not very probable that five men of genius would have died within five years at the University of Göttingen.... A professorial robe is not necessarily a certificate of genius; there may be even at Göttingen some chairs occupied by not very remarkable men.[2]

It is amusing to see Broca explaining away, correction by correction, a reported 100-gram superiority of the brains of Germans over Frenchmen. When, despite his best efforts, Broca found some measurements placing blacks higher than whites, he decided that, after all, those measurements were of no interest. And on it goes. The "objective facts" of science turn out, over and over again, to be the cooked, massaged, finagled creations of ideologues determined to substantiate their prejudices with numbers.

In his debunking of the "data" of anthropometry, Gould follows the model set by Leon Kamin's brilliant muckraking in the byre of IQ studies,[3] but with somewhat different conclusions about the nature of scientific inquiry. Science, he argues, is a social activity, reflecting the reigning ideology of the society in which it is carried out, the political exigencies of the time, and the personal

---

2. Paul Broca, *Bulletin Société d'Anthropologie* 2 (Paris, 1861), pp. 139–207 (quoted by Gould).

3. Leon Kamin, *The Science and Politics of IQ* (Erlbaum, 1974).

prejudices of its practitioners. Racist scientists produce racist science. It is not that they deliberately falsify nature, but that their unconscious prejudices lead them to largely unconscious biases in their methods and analyses, biases that bring them to comfortable conclusions. There are, after all, many ways of explaining observations. How are we to decide among them, except in the light of unspoken assumptions and predispositions?

Like Kamin, I am, myself, rather more harsh in my view of the matter. Scientists, like others, sometimes tell deliberate lies because they believe that small lies can serve big truths. How else are we to understand the doctored photographs discovered by Gould in the report by the American psychologist Henry Goddard on the pseudonymous Kallikak family whose good (*kalos*) and bad (*kakkos*) branches were the living counterparts of the Rougons-Macquarts?

For his part, Sir Cyril Burt, perhaps the most influential psychologist of the twentieth century, *knew* that intelligence was almost perfectly determined by the genes and he was quite willing to make up the data to prove it to people who needed that sort of thing. (His most notorious fabrication was aimed to show that identical twins brought up separately would still be of equal "intelligence.") Burt may indeed have been, as Gould says, "a sick and tortured man" during the last years of his life, but even his biographer, Professor Hearnshaw, admits that Burt was none too scrupulous about numbers at any time.[4] Whether deliberately or

---

4. L. S. Hearnshaw, *Cyril Burt: Psychologist* (Cornell University Press, 1979).

not, there is no evidence that scientists are falsifying nature any less in the twentieth century than they did in the nineteenth.

By the beginning of the twentieth century, the belief that great men had big heads and great criminals big noses had pretty much disappeared from the scientific scene, although it was still part of popular consciousness. When Agatha Christie's young Tommy sees a communist trade-union agitator for the first time, he observes that the fellow

> was obviously of the very dregs of society. The low beetling brows, and the criminal jaw, the bestiality of the whole countenance, were new to the young man, though he was a type that Scotland Yard would have recognized at a glance.[5]

In place of measurements of skull and limb, biological determinist science began to measure intelligence itself. The IQ test, created by the French psychologist Alfred Binet in 1905 as a diagnostic instrument to help teachers help children, became, in the hands of its English-speaking adaptors, Henry Goddard, Lewis Terman, and Charles Spearman, an instrument for arraying everyone along a single scale of mental ability.

Much of the history of the political use of IQ testing in America, especially in helping to justify the Immigration Act of 1924, has been recounted by Kamin,

---

5. Agatha Christie, *The Secret Adversary* (Dodd, Mead, 1922).

who demolished the "data" purporting to show the heritability of IQ differences. Unfortunately, the story of the Cyril Burt frauds is nowhere told in its full richness. Even the summary by Kamin in the book containing his "debate" with H. J. Eysenck[6] is too brief to provide the excitement of psychology's Watergate, which had its own Woodward and Bernstein (Kamin and Oliver Gillie), its outraged denials by Burt's supporters, and its final days of capitulation in the face of the overwhelming evidence of wholesale fakery. And Gould has other fish to fry. *The Mismeasure of Man* looks beyond the politics, the data, and the frauds to address the central epistemological issue about intelligence: "Is there anything to be measured?"

IQ tests vary considerably in form and content. Some are oral, some written, some individual, some given in groups, some verbal, some purely symbolic. Most combine elements of vocabulary, numerical reasoning, analogical reasoning, and pattern recognition. Some are filled with specific and overt cultural referents: children are asked to identify characters from literature ("Who was Mr. Micawber?"); they are asked to make class judgments ("Which of the five persons below is most like a carpenter, plumber, and bricklayer? 1) postman,

---

6. H. J. Eysenck versus Leon Kamin, *The Intelligence Controversy* (John Wiley, 1981). While billed as a debate, this book in fact consists of two independent summary pieces on IQ, followed by brief rejoinders. Eysenck, formerly one of Burt's strongest supporters, here casts his vote for impeachment but says it doesn't matter because the rest of the data on the heritability of IQ is so good. This has become the standard way of handling the Burt frauds, since the facts can no longer be denied.

2) lawyer, 3) truck driver, 4) doctor, 5) painter"); they are asked to judge socially acceptable behavior ("What should you do when you notice you will be late to school?"); they are asked to judge social stereotypes ("Which is prettier?" when given the choice between a girl with some Negroid features and another with a doll-like European face); they are asked to define obscure words (sudorific, homunculus, parterre).

Moreover, the circumstances of testing are laden with tensions. Gould, after reviewing the content of the army classification tests of the First World War, describes at length the intimidating and alien atmosphere in which the tests were given. Complex commands were given just once, in a military style, in English to men many of whom were recent immigrants and some of whom had never before held a pencil. When Gould gave the army beta test, designed for illiterates, in the prescribed style to his Harvard undergraduates, sixteen out of fifty-three got only a B and six got a C, marking borderline intelligence.

The claim is made by their supporters that IQ tests measure a single underlying innate thing, general intelligence, which itself does not develop during the lifetime of the individual but is a cause of the individual's changing overt behavior. In the jargon of educational psychology, "fluid" intelligence becomes "crystallized" by education. Intelligence, so viewed, is not what is learned but the ability to learn, a fixed feature immanent to different degrees in every fertilized egg.

The evidence that there is a unitary intellectual ability is that the results of different tests and of different parts of the same test are correlated with each other. Children who do well on pattern recognition tend to do well in numerical reasoning, analogical reasoning, and so on. But the claim is spurious. IQ tests, like books, are commodities that can yield immense profits for their publishers and authors if they are widely adopted by school systems. A chief selling point of new tests, as announced in their advertising, is their excellent agreement with the original Stanford-Binet test. They have been carefully cut to fit.

Moreover, the agreement of the results of various parts of the same tests has also been built into them. In order for the original Stanford-Binet test to have won credibility as an *intelligence* test, it necessarily had to order children in conformity with the a priori judgment of psychologists and teachers about what they thought indicated intelligence. No one will use an "intelligence" test that gives highest marks to those children everyone "knows" to be stupid. During the construction of the tests, questions that were poorly correlated with others were dropped, since they clearly did not measure "intelligence," until a maximally consistent set was found. The claim that something real is then measured by these selected questions is a classic case of reification. It is rather like claiming, as a proof of the existence of God, that he is mentioned in all the books of the Bible.

A good deal of *The Mismeasure of Man* is taken up with a lucid explanation of the abstruse statistical method used by mental testers to extract a single

16 dimension, g, that is supposed to measure general intelligence. This method, factor analysis, takes a collection of different measurements and combines them into a single weighted average, where the weights are derived from the observed correlations between the measurements. The error, as explained by Gould, is not in the arithmetic, but in the supposition that, having gone through the mathematical process, one has produced a real object, or at least a number that characterizes one. As Gould points out, the price of gasoline was well correlated with the distance of the earth from Halley's comet in recent years, but that does not mean that some numerical combination of the two values measures something real that is their common cause. Even with Gould's help, the reader may remain mystified. The very complexity of the statistical manipulation is part of the mystique of intelligence testing, validating it by making it inaccessible to nonexperts. After all, look how complicated quantum mechanics is, and you can use it to blow up the world.

Gould's view of the biological determinists is that they are doubly blinded: first, by their own racial and ethnic prejudices, and second, by what Gould calls "Burt's real error," the vulgar reductionism that leads them to reify an abstract statistical entity. Yet the analysis is somehow incomplete. With its emphasis on the racism of individual scientists, and on their epistemological naiveté, *The Mismeasure of Man* remains a curiously unpolitical and unphilosophical book. Morton, Broca, Lombroso, Goddard, Spearman, and Burt make their appearance as if from a closet, and smelling a bit of mothballs. They are "men of their time," displaying antique social prejudices which on occasion come back to haunt us in the form of "criminal chromosomes" and a brief eruption of Jensenism. Their biological determinism appears as a disarticulated cultural artifact, nasty and curious, like cannibalism, but not integrated into any structure of social relations.

Biological determinism is the conjunction of political necessity with an ideologically formed view of nature, both of which arise out of the bourgeois revolutions of the seventeenth and eighteenth centuries. These revolutions were made with the slogans "Liberty, equality, fraternity" and "All men are created equal." They meant literally "all *men*," since women were excluded from social power, but they did not mean "*all* men," since slavery and property qualifications continued well into the nineteenth century. Still, one can hardly make a revolution with the cry "Liberty and equality for some!" The problem for bourgeois society (and for socialist society, as well) is to reconcile the ideology of equality with the manifest inequality of status, wealth, and power, a problem that did not exist in the bad old days of *Dei Gratia*. The solution to that problem has been to put a new gloss on the idea of equality, one that distinguishes *artificial* inequalities which characterized the *ancien régime* from the *natural* inequalities which mark the meritocratic society. As the Harvard psychologist Richard Herrnstein puts it:

> The privileged classes of the past were probably not much superior biologically to the downtrodden,

which is why revolution had a fair chance of success. By removing artificial barriers between classes, society has encouraged the creation of biological barriers. When people can take their natural level in society, the upper classes will, by definition, have greater capacity than the lower.[7]

Equality then becomes equality of opportunity, and those who fail do so because they lack intrinsic merit. But if we truly live in a meritocratic society, how do we account for the obvious passage of social power from parent to offspring? It must be that intrinsic merit is passed in the genes. The doctrine of grace is replaced by the Laws of Mendel.

The emphasis in *The Mismeasure of Man* on racism and ethnocentrism in the study of abilities is an American bias. IQ testing was widespread in France long before there were significant numbers of Algerians there, and Sir Cyril Burt's most influential educational invention, the British eleven-plus exam, long antedated the influx of West Indians and Pakistanis. Lombroso's criminal anthropology had nothing to do with race and ethnicity, but with the same *classes laborieuses, classes dangereuses* that concerned Eugène Sue. In America, race, ethnicity, and class are so confounded, and the reality of social class so firmly denied, that it is easy to lose sight of the general setting of class conflict out

---

7. Richard Herrnstein, *IQ in the Meritocracy* (Atlantic/Little, Brown, 1973), p. 221.

of which biological determinism arose. Biological determinism, both in its literary and scientific forms, is part of the legitimating ideology of our society, the solution offered to our deepest social mystery, the analgesic for our most recurrent social pain. In the words of Charles Darwin, quoted on the title page of *The Mismeasure of Man*, "If the misery of our poor be caused not by the laws of nature, but by our institutions, great is our sin."

The disarticulation of social relations, the alienation of man from land, the creation of what C. B. MacPherson calls "possessive individualism,"[8] began in the fourteenth century with the market-town corporations, and slowly became the dominant mode of our society. They brought with them an alienation and objectification of nature. The natural world was seen less and less as an organic unity, an extension of the Mind of God. Like the body social, the body natural came to be an assemblage of elements, interacting with each other, yet each possessing its intrinsic and independent properties. No longer do we "murder to dissect," but rather do we expect to discover the true nature of the world by taking it to bits, the bits of which it is truly made. In this sense Descartes was as much a founding father of our society as Paine or Jefferson.

It is easy to criticize the vulgar materialism of Spearman and Burt, who thought of intelligence sometimes as a form of elementary energy, sometimes as a liquid

---

8. C. B. MacPherson, *The Political Theory of Possessive Individualism* (Oxford University Press, 1962).

that could be crystallized, but it is not clear that anything else could be expected from them. The reification of intelligence by mental testers may be an error, but it is an error that is deeply built into the atomistic system of Cartesian explanation that characterizes all of our natural science. It is not easy, given the analytic mode of science, to replace the clockwork mind with something less silly. Updating the metaphor by changing clocks into computers has got us nowhere. The wholesale rejection of analysis in favor of an obscurantist holism has been worse. Imprisoned by our Cartesianism, we do not know how to think about thinking.

An Exchange

The exchange that follows was published in the February 4, 1982, issue of The New York Review.

TERRY TOMKOW and ROBERT M. MARTIN, of Dalhousie University in Halifax, Nova Scotia, write:

Everyone will acknowledge that the heritability of intelligence and the reliability of IQ tests raise difficult empirical questions and that the issue is complicated by an unhappy history of prejudice, bad science, and, sometimes, outright fraud. But Richard Lewontin thinks it's worse than that. In his review of Stephen Jay Gould's *The Mismeasure of Man* he adds the surprising claim that anyone who thinks you *could* measure intelligence or examine its etiology is guilty of a *conceptual error*:

> . . . there is the conceptual error. Intelligence, acquisitiveness, moral rectitude are not *things* . . . but mental constructs, historically and culturally contingent. The attempt to find their physical site in the brain and to measure them is like an attempt to map Valhalla. It is pure reification, the conversion of abstract ideas into things. . . . While there may be genes for the shape of our heads, there cannot be any for the shape of our ideas.

As philosophers we warm to the prospect of squelching a raucous scientific controversy from the comfort of our armchairs. But all the conceptual errors here seem to be Lewontin's.

Lewontin never makes clear what "reification" is or why it's a bad thing. His example is poorly chosen. The man who wants to map Valhalla is guilty of a factual, not a conceptual, error; he's mistaken in thinking the place exists. The emphasis on "*things*" suggests that we are being warned off the error of thinking that intelligence is a physical object . . . like a rock or a kidney. That *would* be a conceptual error, but not one that anyone has ever been guilty of. Intelligence is (if anything) a *property* of things (people). But Broca, the IQ testers, and everyone else knew *that*. The question is what is wrong *in principle* with thinking that this property might be measured, inherited, or correlated with special features of the brain?

At one point Lewontin complains that the only evidence for the adequacy of IQ tests is that their results agree with one another. But this just ignores the fact that the tests rank people in an order which corresponds to what we independently judge to be their relative intelligence. Lewontin grudgingly acknowledges this but talks as if this only made the whole business suspect:

> In order for the original Stanford-Binet test to have won credibility as an *intelligence* test, it necessarily had to order children in conformity with the a priori judgment of psychologists and teachers about what they thought indicated intelligence. No one will use an "intelligence" test

that gives highest marks to those children everyone "knows" to be stupid. During the construction of the tests, questions that were poorly
correlated with others were dropped, since they
clearly did not measure "intelligence," until a
maximally consistent set was found.

But this (minus all those sneer quotes) sounds exactly
the right procedure for developing a valid test for
intelligence. We accept the patch and sputum tests for
tuberculosis because their results agree with each
other and with physicians' "a priori" diagnoses of
tuberculosis. Would Lewontin say of these tests:

> The claim that something real is then measured . . .
> is a classic case of reification. It is rather like
> claiming, as proof of the existence of God, that
> he is mentioned in all the books of the Bible.

If Lewontin had said that the psychologists and teachers involved couldn't tell the difference between smart
and stupid people in the first place, he would have a
substantive (but unsubstantiated) criticism. But this
would be the charge of bad judgment, not of "conceptual error."

Lewontin seems to regard the "abstruse" statistical
methods used in drawing up IQ tests as dangerously
highfalutin, but says the real problem

> . . . is not in the arithmetic, but in the supposition
> that, having gone through the mathematical

process, one has produced a real object, or at
least a number that characterizes one.

But this talk about "real objects" is unhelpful. IQ tests
give us numbers which correlate with and predict
some people's rankings of subjects by intelligence.
Where is the mistake? Well . . .

As Gould points out, the price of gasoline was
well correlated with the distance of the earth
from Halley's comet in recent years, but that
does not mean that some numerical combination
of the two values measures something real that is
their common cause.

If Lewontin is admitting here that there is a correlation
between IQ and intelligence, but suggesting that it's
only an accidental correlation, he is conceding a good
deal. An intelligence test that works only by accident
is still a test of intelligence that works. So long as the
comet's distance is reliably correlated with gasoline
prices we can use one figure to determine the other.
What makes this sound odd, of course, is the idea that
the correlation between IQ and (psychologists' assess-
ments of) intelligence could be just a coincidence. That
is unlikely and that is just why we suspect a common
cause. In the same way the fact that psychics *can't* agree
on the color of people's auras is evidence that there is no
such thing as an "aura"; while the fact that independent
oenologists tend to coincide in their rankings of vintages
is evidence that their judgments have a real object.

Mixed up in this are Lewontin's cavils about the "historically and culturally contingent" nature of our judgments of intelligence. The thought here is, of course, that what gets counted as "intelligence" will vary from culture to culture and from time to time, in the way that, say, judgments of physical beauty vary historically and culturally. Whether judgments of intelligence are relative is an interesting empirical question, not one that can be settled by armchair theorizing or literary anecdotage. But supposing they are, so what? Lewontin seems to think that this shows that what the tests measure (and maybe intelligence itself) is "unreal." Non sequitur, but then Lewontin isn't the only one to think that as soon as "cultural relativity" rears its ugly head, science goes out the window. This is a mistake, as the following fable may help to show.

Suppose that we undertook to produce "Handsome Tests." Funded by grants from the NSF and Elizabeth Taylor we set about finding a set of physical measures which distinguish good-looking men from the rest of us. We begin with a large sample of men and ask women to rank them by looks. We take careful measurements of the men's physical attributes, apply "abstruse statistics" and come up with the h factor—a complex ratio of nose length to shoe size—that tests out well in predicting who will be regarded as better looking than whom. Success in finding a reliable Handsome Quotient would depend upon and *be evidence for* there being some *real* feature in common among men judged to be handsome. Determining

HQ's would be useful in all sorts of ways, e.g., we could use it as a tool for resolving the nature-nurture question with respect to handsome ("tighter shoes will help but you're stuck with that nose"). But now suppose we notice that the women who produced our original target ranking were all Americans and that their judgments are wildly at odds with those of Russian women. Obviously we can no longer claim that *h* measures handsome *tout court*; we'll have to say that it tests for what counts as handsome among American women. But three things are worth noting here. First, despite the relativity we haven't stopped measuring something *real*. The difference between being good-looking and not—if only to Americans—is nothing to sneeze at; being good-looking to the women around you confers a *biological* advantage. Second, the relativity doesn't mean that our test is useless for a general science of man. The test would be an important first step in producing a general theory of the cultural determinates of such judgments. Third, and most important: *however* much verdicts on handsome vary from culture to culture, it can still be the case that these judgments turn on characteristics that are *biologically* determined. Though assessed differently by different cultures, physiognomy is the primary determinate of good looks and—plastic surgery aside—physiognomy is largely a matter of heredity. Even a "historically and culturally contingent" quality can be "shaped by our genes."

STEVEN ORZACK, of the Museum of Comparative
Zoology at Harvard University, writes:

In his recent review of Steven Jay Gould's new book
Richard Lewontin makes the claim that present-day
scientists are "imprisoned" by "the atomistic system
of Cartesian explanation that characterizes all of our
natural science." I wish to point out that this statement represents an oversimplified view of a complex
situation.

   We are told that scientists are imprisoned by their
reductionism and that reductionism has failed to solve
many important problems in "natural science."
Indeed it has, but Lewontin should realize that failed
attempts are not the same as failed research strategies.
For years there were failed reductionist attempts at
explaining the detailed structure of the atom yet understanding was eventually achieved via reductionist
means. Similarly, failed attempts at explaining the
general patterns of animal development (such as
Spemann's organizers) do not necessarily invalidate
the reductionist effort as such.

Classical reductionism is the belief that the properties
of a system at one level are wholly explainable in
terms of the properties of the components present at a
"lower" level. The whole is merely the sum of its
parts. Yet this mode of scientific analysis may be quite
rare in its pure form. The physicist trying to determine
the equilibrium behavior of gas molecules cares not
a bit (at least now) for the "strangeness" of the

subatomic particles within. Nor do I as a geneticist always attempt to explain genetic processes simply in terms of the "basic" components of the system. For example, in my work on the genetic basis of sex ratio variation in an insect, my collaborator and I have identified various genetic and environmental determinants of sex ratio differences. The careful reader will note that the previous statement contains a bit of reductionism: the distinction between "genotype" and "environment." But one must start *somewhere*. Lewontin would have us believe that most scientists stop thinking after decomposing the system into parts. Not only is this not true (see below) but it is wrong to think that such a decomposition is incompatible with a recognition that a system has an interactive nature to it. My collaborator and I have recognized that certain environments change the expression of sex ratio genes in unique ways. We recognize a simple interaction. This sort of analysis is not uncommon because decomposing the system may be the only way of determining *why* the whole is more than the sum of its parts. Modern natural science is more pluralistic in its methods than Lewontin's statement would imply.

Of course, it might be asserted that the type of scientific research strategy described above has not been used in the past when great scientific (say, biological) discoveries were made. Were Avery, Beadle, Darwin, Morgan, Pasteur, or Wright imprisoned by their reductionism? A partial answer to this question can be found in a recent and fascinating scientific biography of Thomas Hunt Morgan. Morgan is the person who provided, along with his co-workers, much of the foundations of modern genetics. Surely *he* would be a reductionist. Yet Garland Allen, the author of this biography, argues that Morgan

> ... recognized the importance of studying com-
> plex processes initially by breaking them down
> into their component parts, but he did not
> believe that every biological problem could find
> its only satisfactory explanation in purely physi-
> cal or chemical terms. Physics and chemistry
> were helpful in understanding biological prob-
> lems, but an organism was something more than a
> "bag of molecules."[9]

Allen even explicitly characterizes Morgan as a "dialectical materialist." This is an arguable point, but it is clear that Morgan was quite able to make great discoveries using his approach to the study of biologi-cal phenomena. His approach is common.

Of course, tremendous effort will be required to solve the many important biological problems remain-ing. A few new facts will not allow us to "understand" the brain, for example. Nevertheless, there is no rea-son to believe that the varied methods of modern nat-ural science will not allow us to eventually achieve such an understanding.

It is then a straw man that Lewontin creates when

---

9. G. E. Allen, *Thomas Hunt Morgan* (Princeton University Press, 1978), p. 328.

30

he writes of a reductionism which imprisons.[10] Indeed, his "Cartesianism" is a model of scientific inquiry of which it has been said there are no cases.[11]

RICHARD LEWONTIN replies:

I take it as a severe criticism of my ability to write that two professors of philosophy can have so misread my explication of Gould's book. I will try to make amends by explaining the matter again.

The height of a person is a natural attribute of a real object. If I average the heights of ten people, that average is not an attribute of any real object. There is no person with such a height, nor does it characterize the height of the collection of individuals since a collection of people does not have a height. The average is not even a height. It is simply the sum of a lot of measurements divided by the number of measurements. It is a mental construction. To assert that it is a real attribute of a real thing is an act of reification (indeed, double reification!). Again, if I multiply a person's height by the number of letters in her name and divide that by the zip code of her residence, I will get an index that may do quite a good job of picking her out from a crowd.

10. What may be imprisoning is philosophy of science which uses oversimplified models to explain the dynamics of the scientific process.

11. See W. C. Wimsatt, "Reductionistic Research Strategies and Their Biases in the Units of Selection Controversy," in *Scientific Discovery: Case Studies*, edited by Thomas Nickles (D. Reidel Publishing Co., 1980). See especially page 216 and following.

It is not, however, the characterization of a physical attribute, and to claim that it was would be an act of reification. Finally, I may ask a lot of children questions about language, geometrical patterns, numbers, and social attitudes, then construct the matrix of population correlations among these different sets of questions, rotate the axes of the matrix, and find its principal eigenvector, call the vector "g," project an individual's scores onto the principal eigenvector, and come up with a single characterization of that individual.

That abstruse calculation may do a moderately successful job of picking out children whom people believe to be "intelligent" (or handsome) and may even be of a little value in predicting who will make more money as an adult (parents' income and occupation are much better predictors), but that does not make g a natural attribute. To claim that g is a natural attribute is to reify a mental construct. When Spearman and Burt went from constructing the value g to asserting that it measured a physical property, intelligence, that could be "fluid" or "crystallized" and that was a form of energy, they were not making a factual error, but a conceptual one. As to Valhalla, I would rather not get embroiled in the historical issue of whether it was a false hypothesis or a pure mental construct. If we read "New Jerusalem" for "Valhalla," the point is made without ambiguity.

The second misunderstanding is the question of the correlation between IQ scores and so-called "intelligence." Certainly IQ scores accord with a priori judgments about who is intelligent. That is because the

Stanford-Binet test was cut and fit until it picked out such people. The issue that Gould and I were addressing was whether the *agreement* in the results of different IQ tests and parts of IQ tests could be taken as evidence that they were measuring something real. That agreement is not evidence because tests are not independent of each other but are adjusted to agree with the Stanford-Binet. Tomkow and Martin have thoroughly muddled "intelligence" with notions about intelligence. IQ tests do pick out people whom teachers and psychologists think are intelligent. Unfortunately, that fact has confused even our philosophers into thinking that the tests pick out people who have a physical, heritable, internal property, "intelligence," that stands apart from socially determined mental constructs. That confusion is enshrined in E. G. Boring's famous definition of intelligence as what IQ tests measure. The Catholic Church has a very elaborate, exacting, and successful test procedure, including the attestation of miracles, for finding out people whom its members regard as being "saintly." But saintliness remains a mental construct, just like intelligence. It is not simply our "judgments of intelligence" but the very idea of intelligence that is a historically contingent mental construct.

It is important to point out that the distinction between mental constructs and natural attributes is more than a philosophical quibble, even when those constructs are based on physical measurements. Averages are not inherited; they are not subject to natural selection; they are not physical causes of any events. There are no "genes for handsomeness" or "genes for intelligence" any more than there are "genes for saintliness." To assert that there are such genes is a conceptual, not a factual, error and one that has major consequences for scientific practice and social analysis.

Orzack's point, about the failure of Cartesian analysis, comes down to a difference of opinion. It is, of course, dangerous to claim that the brain and the embryo will never be understood using our present concepts. It is the great irony of molecular biology that, inspired by Schrödinger's *What Is Life*, it began with the belief that the ordinary laws of physics would not suffice to explain biological phenomena, and ended up with a description of basic hereditary processes that looks for all the world like a Ford assembly plant. Nevertheless, it is very unlikely that we are waiting for just a few new facts or experimental techniques to crack the problem of the central nervous system. Questions about the brain combine direct physical properties with metaphysical constructs that we cannot seem to avoid. It is a very different thing to ask "What are genes made of?" than to ask "What is the anatomical and molecular basis of thinking?" The first is well within the framework of Cartesian analysis while the second has that nasty word "thinking" in it. The problem is to bridge the gap between substance and thought, to do in conscious language what our brains do by their very nature.

Epilogue

34

The struggle over biological determinism never ceases. In the nineteen years since the publication of *The Mismeasure of Man*, there have been repeated claims that differences in cognitive ability and personality between individuals, social classes, sexes, or races are genetically determined. Only the emphasis has shifted. Successful political and social agitation against crude racism has made assertions of the intrinsic psychic superiority of one race over another totally unacceptable in respectable circles. The only academic to have tried this tack lately, a professor of psychology at York University, botched the job rather badly.[12] His claim was that Africans are duller but sexier than Europeans and Asians because evolution in a tropic clime endowed them with smaller brains and larger genitalia. Unfortunately for his thesis, the "data" were an unculled and unevaluated collection of reports, including one from 1865 on "5 Negro men" and one in 1874 on "10 Japanese executed by decapitation," and he got his evolutionary theory precisely upside down. On the other hand it is still within respectable limits to assert that genetically determined differences between the sexes account for the greater social power of men, if not their greater ability in algebraic topology. Thus, there continues to be a literature, both feminist and antifeminist, asserting

12. J. Philippe Rushton, *Race, Evolution and Behavior: A Life History Perspective* (Transaction, 1995).

a direct causal role of the sex chromosomes in shaping personality and mental abilities, and a counterliterature aimed at exposing the factual and logical errors of that determinist position. The continued life of biologistic explanations of the sexual division of social power, in contrast to the disappearance of such arguments from discussions of race, owes a good deal to those feminist theorists who insist on the intrinsic determination of a distinct feminine worldview.[13]

While no educated person now wants to admit to a stark racism and only a hardy few will publicly insist that women really can't do higher mathematics, the claim that social class differences are coded in the genes continues to have a legitimacy reinforced by public intellectuals. After all, in a society of equal opportunity and social mobility, how is one to explain the differences in social power, if not by differences in intrinsic ability? That is, social class differences are explained as the consequence of the sorting out of individual differences and, surely, no one can find the claim to be unreasonable that individuals differ innately.

It is on this plane of individual differences and the social sorting consequent on them that the main struggle around biological determinism continues. Its most famous literary incident was the publication in 1994 of Richard Herrnstein and Charles Murray's *The Bell Curve: Intelligence and Class Structure in American Life*. Their argument for the genetic superiority of those in

---

13. For a detailed discussion of this aspect of the struggle over biological determinism, see Chapter 6.

the 39.6 percent tax bracket followed closely the structure of Herrnstein's earlier book, *IQ in the Meritocracy*: American society is a meritocracy with high, although not perfect, social mobility; success depends on individual cognitive ability; cognitive ability is measured by IQ tests; IQ test scores are highly heritable. Hence, differential social power and status are a consequence of different genetic endowments. What had changed in the intervening twenty years was that there were new studies on heritability of IQ.

It is usually supposed that the best way to estimate the influence of environment as opposed to genes on human traits would come from a comparison of identical twins raised apart, because the confounding effect of environmental similarity for twins raised together can be eliminated, leaving only their genetic identity as the cause of their similarity. By the time *The Mismeasure of Man* was published, however, identical-twin studies had been discredited, partly because the samples were so small and the experimental protocols were so bad (twins raised "apart" turned out, for example, to live in the same village, having been raised by friends or relatives of the biological parents), and partly because of the discovery of the outright fraud committed in the "studies" by the leader in the field, Cyril Burt.[14] Then, in 1990, an article appeared in *Science*, the leading American general scientific journal, reporting the results of the Minnesota Study of

---

14. For an analysis of this literature and the first indications of the fraud, see Kamin, *The Science and Politics of IQ*.

Twins Raised Apart.[15] The sample was of reasonable size, twin pairs raised apart were pursued until they agreed to be in the study, the diagnosis of twins as being identical was made by biochemical and fingerprint similarities, several tests of intelligence were used and marked by different observers, home environments were scored numerically by yet other observers, the lengths and ages of contact between twins were accounted for. The authors seemed to have self-consciously covered all the bases and they estimated a heritability of 75 percent, higher even than the previous flawed or fraudulent estimates. The Minnesota twin study has proved a great boon to the supporters of genetic determinism.

The issue, however, is not one of sample size or of objective testing. The gross errors and frauds associated with previous twin studies made them easy targets for devastating criticisms on the simplest level of how experiments are to be done. As a result there was little attention paid to the real conceptual errors. First, despite the everyday understanding of the word, "heritability" is not a measure of how much a trait can be altered by environmental change. A trait can be 100 percent heritable in the circumstances in which that heritability was measured, yet be easily changed. If that were not true, medical genetics would lose most of its interest. People with two copies of the mutation for Wilson's disease used to die in adolescence or early

15. T. J. Bouchard, D. T. Lykken, M. McGue, N. L. Segal, and A. Tellegen, "Sources of Human Psychological Differences: The Minnesota Study of Twins Raised Apart," *Science*, No. 250 (1990), pp. 223–250.

adulthood with absolute certainty, because of the lack of a single enzyme. Now they survive by taking a simple pill that makes up for their chemical deficiency. Wilson's disease used to be 100 percent heritable, but is no longer. The case of Wilson's disease, or other simple metabolic disorders, also illustrates the importance of specifying quite precisely what trait is said to be heritable. The heritability of the absence or presence of the enzyme in this case is still 100 percent. There is no environment in which people with two copies of the Wilson's disease mutation will produce the enzyme. But the heritability of the disease, as a disease leading to adolescent death, is now zero, at least for any population with easy access to the treatment. The "heritability" of a trait only measures the proportion of variation among people that is caused by the variation of their genes in the present array of environments and for that specific trait. Thus, an estimate of the heritability of a characteristic has no predictive or programmatic value. Its heritability under present circumstances contains no information about the trait in future (or past) circumstances, nor is it of any use for designing programs of intervention. Were God to appear to me in a dream telling me the heritability of, say, coronary artery disease, to the fourth decimal place, I could not use that information for any program of amelioration, prevention, or cure, because it would tell me nothing useful about the pathways of mediation. To say that genes are somehow influential is to say nothing, because genes are somehow influential in all traits of all organisms. What is required for a program of alteration of a trait is an understanding of its actual mediation. Wilson's disease is preventable because we have a detailed knowledge of the metabolic pathway that has been disrupted, and nothing has been added by the knowledge that the disruption is a consequence of a genetic mutation.

Second, the array of environments in which separated identical twins are raised may miss the essential causal social variables. Before we know what to measure about the family environment we have to have a theory of what matters. That is, we bring to any such study a prior theory of the social and individual determinants of psychic development.[16] Suppose, for example, that the self-confidence of parents were the dominant influence on IQ test performance and that, for some reason, identical twins raised apart were almost always put into families that had a greater than average sense of control of their environment. Then the variation in test scores among these twin pairs would be unrepresentative of the general population in several ways. They would have higher IQ scores than the general average and there would be less than normal variation among them because they had been put into a restricted group of families.

If this entirely made-up theory seems fanciful, it should be pointed out that the Minnesota twins had an average IQ score eight points higher than the population

---

16. See Chapter 7 for a discussion of the way in which prior social theory influences both the questions that are asked and the variables that are analyzed in social scientific research.

average and a variance in scores only half as large as the general population. The consequence is that 75 percent of them had a higher score than the general average. The Minnesota investigators say nothing about this discrepancy. Moreover, in the absence of an adequate theory of determination, the family environmental factors that were measured by the investigators, and whose correlations with IQ were calculated, would be irrelevant to the problem. Finally, no matter what the heritability of IQ test performance, the argument of *The Bell Curve* depends on the undemonstrated claim that intelligence, as defined by IQ score, is an important determinant of social success and on the untrue belief in unrestricted social mobility.[17]

In the end, despite all the effort to make the study of social structures "objective," we always bring to the study of society an already formed social theory.

---

17. See, for example, the classic study of social mobility by P. Blau and O.D. Duncan, *The American Occupational Structure* (John Wiley, 1967). The rate of movement between white- and blue-collar categories must also take into account that this movement is almost always horizontal, with blue-collar production workers being turned into white-collar office workers and retail clerks at equal or lower pay levels and less employment security.

Chapter 2

Darwin's Revolution

"Darwin's Revolution" was first published in
The New York Review of Books *of June 16, 1983,*
as a review of Darwin for Beginners, *by Jonathan*
Miller and Borin Van Loon (Pantheon, 1982);
Evolution Now: A Century After Darwin, *edited by*
John Maynard Smith (W. H. Freeman, 1982);
Evolution Without Evidence: Charles Darwin and
"The Origin of Species," *by Barry G. Gale (Uni-*
versity of New Mexico Press, 1982); The Monkey
Puzzle: Reshaping the Evolutionary Tree, *by John*
Gribbin and Jeremy Cherfas (Pantheon, 1982);
The Myths of Human Evolution, *by Niles Eldredge*
and Ian Tattersall (Columbia University Press,
1982); Science on Trial: The Case for Evolution,
by Douglas J. Futuyma (Pantheon, 1983); Abusing
Science: The Case Against Creationism, *by Philip*
Kitcher with Patricia Kitcher (MIT Press, 1982);
and Darwinism Defended: A Guide to the Evolution
Controversies, *by Michael Ruse, foreword by Ernst*
Mayr (Addison-Wesley, 1982).

SCIENTISTS ARE INFATUATED with the idea of revolution. Even before the publication of Thomas Kuhn's *The Structure of Scientific Revolutions*,[1] and with ever increasing frequency after it, would-be Lenins of the laboratory have daydreamed about overthrowing the state of their science and establishing a new intellectual order. After all, who, in a social community that places so high a value on originality, wants to be thought of as a mere epigone, carrying out "normal science" in pursuit of a conventional "paradigm"? Those very terms, introduced by Kuhn, reek of dullness and conventionality. Better, as J. B. S. Haldane used to say, to produce something that is "interesting, even if not true." As a consequence, new discoveries are characterized as "revolutions" even when they only confirm and extend the power of ideas that already rule.

So, for example, the discovery, by J. D. Watson and Francis Crick, of the structure of DNA, the stuff of the genes, is often regarded as a scientific revolution. Yet,

---

1. University of Chicago Press, 1962.

44

as Watson himself points out, everyone was waiting for the structure; everyone knew that when it was worked out, an immense variety of phenomena could immediately be fitted in.[2] The model of the organism as a Ford assembly plant was already in place, and the fenders and bumpers were already stockpiled; all that was needed was the key to turn on the assembly line. The discovery of the structure of DNA has been immensely fruitful, for all of present-day molecular biology and genetics was made possible by it, but it has not made us see the biological world in a different way. It has not been upsetting, but fulfilling.

As in politics, so in science, a genuine revolution is not an event but a process. A manifesto may be published, a reigning head may drop into a basket, but the accumulated contradictions of the past do not disappear in an instant. Nor do the supporters of the *ancien régime*. The new view of nature does indeed resolve many of the old problems, but it creates new ones of its own, new contradictions that are different from, but not necessarily any less deep than, the old. And waiting, just across the border, are the intellectual *somocistas*, saying, "I told you so. What did you expect?" trying to convince us that the old way of looking at nature was correct after all. Of course, the old view of nature can never return, but rather new revolutions displace the old ones.

There have been only two real revolutions in biology since the Renaissance. The first was the introduction of

---

2. James D. Watson, *The Double Helix* (Atheneum, 1968).

mechanical biology by William Harvey and René Descartes. While their manifestoes declaring that animals were machines were published early in the seventeenth century,[3] it was not for another 250 years that the mechanistic revolution in biology was fully achieved. The difficulties of the reductionist mechanical view of biology have given prolonged life to vitalism and obscurantist holism, relics of an organic view of nature that comes down to us from the Middle Ages and, at the same time, have driven some biologists to search for yet another conceptual revolution to solve the mysteries of mind and of development.[4]

The second biological revolution, to which we attach the name of Darwin, is still being consolidated. Although its manifesto, *On the Origin of Species*, appeared in 1859, it was not until the 1940s that Darwinism really established a hegemonic hold on such branches of biology as classification, physiology, anatomy, and genetics. It is still under external siege by the restorationist armies of creationism, while at the same time it is undergoing a severe internal struggle to define its own orthodoxy and to resolve its own contradictions. The hundredth anniversary, in 1982, of Darwin's death was marked by an enormous production of books, a triumph of the power of modern capitalism to turn ideas into commodities, equaled only by what is

---

3. Harvey's *Exercitatio anatomica de motu cordis et sanguinis in animalibus* was published in 1628, and Descartes's *Discours* in 1637.

4. I have discussed these problems at length in "The Corpse in the Elevator," *The New York Review*, January 20, 1983, pp. 34–37.

being done to commemorate the death of Marx. The year between April 1882 and March 1883 was a bad one for revolutionaries, but a great opportunity for publishers.

Some books, like *Darwin for Beginners*, are meant to introduce the content and history of Darwinism to the layperson; some, for example *Evolution Now: A Century After Darwin*, to expose for a professional audience the internal state and modern problems of the theory itself; yet others, such as *Evolution Without Evidence*, are part of the immense industry of Darwin scholarship that gives employment to large numbers of historians and philosophers of biology and provides the material for their professional journals. Then there are books such as *The Monkey Puzzle* and *The Myths of Human Evolution* that are concerned with the quest for that mythic pot of paleontological gold, the Missing Link. Most immediately relevant are works such as *Science on Trial* and *Abusing Science* that defend Darwinism against the real besiegers without the walls, or, like *Darwinism Defended*, that see sinister subversives within the very citadel, conspiring to destroy what the barbarian hordes are unable to shake.

Darwin surely was a revolutionary, but there is a certain confusion about what constituted his epistemological break with the past. Clearly it was not the idea of evolution itself. Darwin was rather the inheritor than the creator of the view that life evolved. Indeed, the nineteenth century was a period in which rampant evolutionism became a general worldview, one not restricted to the history of life. Evolutionary cosmology began in 1796 with Laplace's nebular hypothesis for the origin of the solar system. Charles Darwin's grandfather, Erasmus, had postulated the evolution of all organisms from "rudiments of form and sense" in his epic poem *The Temple of Nature*. Chartism and the discoveries of geology conspired to popularize the view that change and instability were universal. Not even nature could be counted on to hold the line:

> "*So careful of type?*" *but no.*
> *From scarpéd cliff and quarried stone*
> *She cries, "A thousand types are gone;*
> *I care for nothing, all shall go.*"

> (TENNYSON, *In Memoriam*, 1844)

Change, ceaseless change, "a beneficent necessity," as Herbert Spencer called it, preoccupied the scientific, literary, philosophical, and political consciousness of European culture from the suppression of the *Encyclopédie* in 1759 to the instantaneous bookshop success of *On the Origin of Species* in 1859. For a revolutionizing bourgeoisie, the only constant was the process of change itself. Their battle cry was already formulated 100 years before Darwin by Diderot in *Le Rêve de d'Alembert*: "*Tout change, tout passe. Il n'y a que le tout qui reste.*"

Although Darwin did not invent the idea of evolution, he certainly was responsible for its widespread acceptance. *On the Origin of Species* not only precipitated the intense popular debate on evolution, but was in itself a convincing argument. Its persuasiveness arose only partly from the assemblage of evidence from natural history and paleontology that evolution had occurred, but largely from the construction of a plausible theory of how it occurred. When we speak of the "theory of evolution," a constant confusion arises between the fact of the historical transformation of organisms over the last three billion years and a detailed and coherent theory of the dynamics of that historical process. There is no disagreement in science about whether evolution has occurred. There is bloody warfare on the question of how it has occurred.

It is this confusion between fact and theory that is exploited by creationists, who use the struggle among scientists about the process to claim that the phenomenon itself is in question. *Science on Trial* and *Abusing Science* both deal with the structure of evolutionary fact and theory as it is confronted by creationist attacks. In *Science on Trial*, Douglas J. Futuyma, a biologist, makes "The Case for Evolution" by a lucid exposition of what is known about the history of organisms and about the processes of inheritance and natural selection, devoting only a single chapter to refuting creationist arguments directly. Philip Kitcher, a philosopher, argues "The Case Against Creationism" by exposing point by point the epistemological errors and willful intellectual dishonesty that characterize creationist claims. Anyone seriously interested in understanding the scientific and philosophical content of the struggle over evolution and creation must read both of these books, Futuyma first and Kitcher second. Anyone who is still confused after doing so has just not been paying attention.

For all their lucidity in dealing with the content of evolutionary theory and the creationist attack upon it, Futuyma and Kitcher give us only academic logic and natural history. They leave us mystified about the origins of the struggle and why people care all that much about it. Why only in America? Why now? Why the passion, commitment, expenditure of time and money by fundamentalists?

Creationism can only be understood as part of the history of southern and southwestern American populism. Earlier in this century, tenants, small holders, and miners shared the perception that their lives were controlled by rich bankers, merchants, and distant absentee corporations who were their creditors and their employers. The same regions of America that were strongest in fundamentalist Christianity were strongest in socialism. Eugene Debs received more votes in 1912 from the poorest counties in Oklahoma, Texas, and Arkansas than in northern urban centers. The best-selling weekly magazine in the United States in 1913, surpassing even the *Saturday Evening Post*, was the socialist *Appeal to Reason* published in Gerard, Kansas. Farmers rode to summer socialist camp meetings in buckboards with red flags flying.[5] If the poor could

---

5. James R. Green, *Grass-Roots Socialism* (Louisiana State University Press, 1978).

have no control over their economic and political lives, at least they could control their cultural and religious lives and what went into the heads of their children. And so they did. As late as 1956, my children in North Carolina learned that "God makes the flowers out of sunshine." Evolution was taught barely at all in the classroom; it was not in the school texts.

Then came the challenge of Soviet science and the world was turned upside down. The National Science Foundation supported scores of professors from eastern establishment universities to rewrite the biology textbooks to bring them up to date, and then saw to it that the school curricula were everywhere revised. Suddenly the intellectual culture of the well-to-do had invaded the homes of ordinary folk in Texas, Oklahoma, and Arkansas, and of the descendants of Okies and Arkies in California. In response, the forces of Christian fundamentalism began to assemble, to prepare the campaign that has only reached full force in the last few years. Though Futuyma and Kitcher speak with the tongues of biologists and philosophers, if they have no historical understanding, their arguments are as sounding brass or a tinkling cymbal. As an implacable but compassionate enemy of religion pointed out:

> The abolition of religion as people's illusory happiness is the demand for their real happiness. The demand to abandon their illusions about their condition is a demand to abandon a condition that requires illusions. The criticism of religion is,

then, in embryo, a criticism of the vale of tears whose halo is religion.[6]

A second difficulty of both Futuyma's analysis and the Kitchers' is that they lose courage on the question of materialist explanations of the world. In the last chapter of *Abusing Science*, the Kitchers offer a Newtonian first cause as a form of religious belief that is not in necessary conflict with natural science. God, according to this view, set the world rolling according to laws of His own invention and has since kept His hands off. The job of natural sciences is then to explicate the divine order. But that analysis misses the essential differences between a God who *has not* intervened (except perhaps to produce an occasional incarnation or resurrection) and a God who *cannot* intervene. Nature is at constant risk before an all-powerful God who at any moment can rupture natural relations. For sufficient reason, He may just decide to stop the sun, even if He hasn't done so yet. Science cannot coexist with such a God. If, on the other hand, God *cannot* intervene, he is not God; he is an irrelevancy. By failing to confront this problem, biologists and philosophers may make Unitarians and agnostics feel that ontological pluralism is a happy solution, but they haven't fooled any fundamentalists, who know better.

If Darwin's revolution was not in proclaiming evolution as a fact, then it must have been in his theory of its

---

6. Karl Marx, *Toward a Critique of Hegel's Theory of the Right* (1844).

mechanism. And what was that theory? Why, "natural selection," of course, which then makes the theory of natural selection the very essence of Darwinism and any doubt about the universal efficacy of natural selection anti-Darwinian. There is a form of vulgar Darwinism, characteristic of the late nineteenth century and rejuvenated in the last ten years, which sees all aspects of the shape, function, and behavior of all organisms as having been molded in exquisite detail by natural selection—the greater survival and reproduction of those organisms whose traits make them "adapted" for the struggle for existence. This Panglossian view is held largely by functional anatomists and comparative physiologists who, after all, make a living by explaining what everything is good for, and by sociobiologists who are self-consciously trying to win immortality by making their own small revolution. Evolutionary geneticists, on the other hand, who have spent the last sixty years in detailed experimental and theoretical analysis of the actual process of evolutionary change, and most epistemologists take a more pluralistic view of the forces driving evolution.

An occasional philosopher has allied himself or herself with the "adaptionists," who give exclusive emphasis to natural selection, and one such, Michael Ruse, makes a characteristic presentation in *Darwinism Defended*. Darwinism, the representative of objective modern science, is under ideologically motivated attack. Professor Ruse is alarmed: "'Darwinism,' as I shall refer to Darwin-inspired evolutionary thought, is threatened from almost every quarter." Well, not from *every* quarter, just the right and left flanks, it seems. First, the fundamentalists, supported by Ronald Reagan, make a know-nothing assault from the right. No sooner have real evolutionists wheeled to face this attack than they are fallen upon by subversive elements from the left, "biologists with Marxist sympathies" and their "fellow travelers" among philosophers who argue "that any evolutionary theory based on Darwinian principles—particularly any theory that sees natural selection as *the* key to evolutionary change—is misleadingly incomplete."

Onto the field, mounted upon his charger perfectly adapted for the purpose, with weapons carefully shaped by selection to spread maximum confusion among the enemy, not to mention innocent civilians, comes Professor Ruse, "trying to rescue . . . from the morass into which so many seem determined to drag them," "Darwin's life and achievements." In all fairness to Professor Ruse, he did not invent this version of events. The theory that evolutionary science is being brutally beaten and cut with crosses, hammers, and sickles made its first appearance in E. O. Wilson's *On Human Nature* as the only plausible explanation he could imagine for the failure of sociobiology to achieve instant, universal, and lasting adherence. The situation of evolutionary theory, however, is rather more complex and more interesting than Professor Ruse's Manichaean analysis suggests.

There are two basic dynamic forms for evolving systems. One is *transformational*, in which the collection of objects evolves because every individual element in the collection undergoes a similar transformation. Stellar evolution is an example. The universe of stars is evolving because every star undergoes the same general set of transformations of mass and temperature during its life cycle from birth to its eventual winking out. The Harvard class of 1950 is getting grayer and flabbier, because each of its members is doing so. Most physical systems and social institutions evolve transformationally, and it was characteristic of pre-Darwinian evolutionary theories that they, too, were transformational. Lamarck held that a species evolved because its individual members, through inner will and striving, changed to meet the demands of the environment.

The alternative evolutionary dynamic, unique as far as we know to the organic world, and uniquely understood by Darwin, is *variational* evolution. In a variational scheme, there is variation of properties among individuals in the ensemble, variation that arises from causes independent of any effect it may have on the individual who possesses it. That is, the variation arises at random with respect to its effect. The collection of individuals evolves by a sorting process in which some variant types persist and reproduce, while others die out. Variational evolution occurs by the change of frequency of different variants, rather than by a set of developmental transformations of every individual. Houseflies, for example, have become resistant to DDT. Because of random mutations of genes that affect the sensitivity of flies to insecticide, some flies were more resistant and some less. When DDT was widely applied, the sensitive flies were killed and their genes were lost, while the resistant forms survived and reproduced, so their genes were passed on to future generations. Thus, the species as a whole became resistant to DDT.

Darwin's problem, and that of anyone trying to produce a theory of evolution, was to explain two apparently distinct features of the organic world, *diversity* and *fit*.

> In considering the Origin of Species, it is quite conceivable that a naturalist . . . might come to the conclusion that species had not been independently created, but had descended like varieties, from other species. Nevertheless, such a conclusion, even if well founded, would be unsatisfactory, until it could be shown how the innumerable species inhabiting this world have been modified, so as to acquire that perfection of structure and coadaptation which justly excites our admiration.[7]

The observations of diversity were strong support for evolution, for the immense variety of species alone seemed to make special creation unreasonable. Once, when J. B. S. Haldane was asked what he could deduce about the Creator from the nature of Creation, he replied, "He must have been inordinately fond of beetles." The marvelous fit of organisms to their environments, however, seemed evidence of a deliberate design. What was so attractive about Darwin's theory

---

7. Charles Darwin, from the introduction, *On the Origin of Species* (1859).

was that it explained both diversity and "that perfection ... which justly excites our admiration" by a single coherent mechanism.

Darwin claimed that it was the sorting process among variants that produced the fit of organisms to the environment. In the "struggle for existence," some variants, because of their particular shapes, behaviors, and physiologies, were more efficient at using resources in short supply, or in escaping predators or other vicissitudes of nature. Thus, the differential survival and reproduction of different variants would be directed by the circumstances of the external world; and so the outcome of the sorting process would bear a close correspondence to that world and its demands. That is, adaptation of the species occurs by sorting among individuals. What vulgar Darwinists fail to understand, however, is that there is an asymmetry in Darwin's scheme. When adaptation is observed, it can be explained by the differential survival and reproduction of variant types being guided and biased by their differential efficiency or resistance to environmental stresses and dangers. But *any* cause of differential survival and reproduction, even when it has nothing to do with the struggle for existence, will result in *some* evolution, just not adaptive evolution.

The Panglossians have confused Darwin's discovery that all adaptation is a consequence of variational evolution with the claim that all variation evolution leads to adaptation. Even if biologists cannot, philosophers are supposed to be able to distinguish between the assertion that "all x is y" and the assertion that "all y is x," and most have. This is not simply a logical question but an empirical one. What evolutionary geneticists and developmental biologists have been doing for the last sixty years is to accumulate a knowledge of a variety of forces that cause the frequency of variant types to change, and that do not fall under the rubric of adaptation by natural selection. These include, to name a few: random fixation of nonadaptive or even of anti-adaptive traits because of limitations of population size and the colonization of new areas by small numbers of founders; the acquisition of traits because the genes influencing them are dragged along on the same chromosome as some totally unrelated gene that is being selected; and developmental side effects of genes that have been selected for some quite different reason.

An example of the last is the redness of our blood. Presumably we have hemoglobin because natural selection favored the acquisition of a molecule that would carry oxygen from our lungs to the rest of our body, and carbon dioxide along the reverse route. That our blood is red, as opposed to, say, green, is an accidental epiphenomenon of hemoglobin's molecular structure, and a few animals, like lobsters, have green blood. This has not stopped adaptationist ideologues from inventing stories about why blood ought to be red, but they are not taken seriously by most biologists.

Despite the fact that mechanisms of nonadaptive evolution are firmly entrenched as part of modern evolutionary explanation and are discussed at some length in Futuyma's book, they are dismissed by Ruse as only "background noise against the main evolutionary tune." It may be, however, that Professor Ruse's ear is not accustomed to counterpoint.

A more realistic and less ideological view of the current problems and controversies in evolutionary theory can be seen in *Evolution Now*, although the lay reader will find it hard going. It includes chapters on the evolution of the structure of genes themselves and how they are grouped together on the chromosome, partly as a consequence of adaptive forces and partly as a trace of purely historical accident. One section concerns the speed of evolution, and whether most evolutionary change may occur during short periods at the time that new species are formed. An ultra-adaptationist section deals with the evolution of behavior and an antiadaptationist section with parasitic DNA. Many of the ideas in this book will turn out to be false alarms, but they illustrate the richness of evolutionary phenomenology as opposed to the poverty of a mindless adaptationist program.

Nothing produced more resistance to Darwin and his supporters than the claim that human beings had themselves evolved from "lower" forms of life. By an ironic inversion, nothing has titillated public interest in evolution in our own time so much as the search for the bones of human ancestors. The surest way to intense, if fleeting, fame and glory is to announce that some tooth, jaw, skullcap, or entire head has just been dug up that is "probably a human ancestor." If it is a half-million years older than the oldest "probable human ancestor" already known, fame and controversy are guaranteed.

There is a problem, however. The only way to be sure that a fossil is really a human ancestor is to find one that is already indubitably human, but then it has no interest. The farther back in time one goes and the greater the differences from us, the more likely it is that the bones belong to some twenty-second cousin twelve times removed. During the last hundred years, there has been a gradual change in the understanding of paleontologists about the shape of evolutionary trees. At one time it was thought that a group of related fossils of increasing ages could be aligned in a single ancestral order or, at most, two or three branches joining the main stem at a few points. It is now reasonably clear that most fossils of different ages cannot be connected in a linear sequence, but represent a small sample from a lot of parallel lines. Evolutionary trees have become bushes. Since fossils, especially of vertebrates, are few and far apart, there are big gaps in the fossil record, and any temptation to arrange fossils in a linear order is likely to be overturned when the next bone is dug up. Often a supposed ancestor will turn out to be contemporaneous or even later than the species to which it supposedly gave rise. Between two million and one million years ago there were four known coeval "apemen," including three that probably used tools.

What makes the situation all the more confused is that the shapes of organisms do not change uniformly in time. There are periods of rapid change and periods of relative stasis. That observation has led to the theory of "punctuated equilibrium," which exaggerates and universalizes that temporal irregularity. According to punctuationists, gaps in the fossil record are real and are the consequence of long periods of absolutely no change in organisms followed by a paltry few thousand years of very rapid evolution. Gradualists, who scornfully refer to this theory as "punc. eq.," say that it is just hard to find the intermediates because suitable fossil-bearing strata are so seldom exposed, so what appears as "punc. eq." is just punk rock.

The theory of punctuated equilibrium is applied to the human fossil record in Niles Eldredge and Ian Tattersall's *The Myths of Human Evolution*. The authors are judicious but biased, minimizing the observed differences between fossils so that they can claim that species show no significant change over a million years or more. There just is not enough material evidence to make a convincing case for punctuated equilibrium from human fossils, but one thing is clear from their analysis. The search for the Missing Link, the oldest form that is clearly in the direct line of human ancestry, is a delusion. No one knows, or ever will know with the sort of evidence upon which we now depend, whether any fossil is a direct ancestor of the people who dig them up and write books about them. That will not stop the claims. One doesn't get many column inches with the announcement that yet another bit of yet another relative of unknown degree has been found in the deserts of East Africa.

Despite the myth of song and story, we did not descend from monkeys or apes, at least not from any forms of them now alive. But we did have common ancestors with chimpanzees and gorillas, not all that long ago. Just how long ago, and who among the living apes is our nearest relative, is not known for certain.

The Monkey Puzzle by John Gribbin and Jeremy Cherfas makes a convincing, if breezy, argument, accessible to a lay reader not frightened by an occasional number, that we are as closely related to the sulky gorilla as to the lovable chimp, and that our genes parted company from theirs only about four to five million years ago. The evidence comes from a form of nonadaptive evolution that has turned out to be one of the most powerful tools biologists have in reconstructing ancestry. It appears that some of the building blocks, the amino acids, of which some of our proteins are made, can be replaced with blocks of slightly different molecular form without affecting the function of the proteins. As the generations succeed one another, this replacement occurs at a clocklike rate, independent of natural selection for specific adaptation. If this clock can be calibrated, by counting the number of replacements that separate two species whose time of evolutionary divergence is known from the fossil record, even approximately, then for other species without a fossil record a time of divergence can be estimated. It is this technique that has shown us to be a mere five million years separated from our common ancestor with Mr. Jiggs.

Unfortunately, Gribbin and Cherfas seem to think that this information tells us something important about the human condition. After all, they say, our proteins are only 1 percent different from those of apes.

But this is a spurious comparison. The authors have forgotten, in their anxiety to say something profound, that the very method they describe with such clarity depends critically on protein differences that have no functional significance in the first place. If calibration of the molecular clock uses nonadaptive evolutionary change as its basis, then how can they expect to draw adaptive meaning from the amount of that change? More generally, how do we convert the percentage difference in molecular composition into a percentage difference in shape, size, or the ability to do biochemistry? I would turn the comparison upside down and remark how little difference in protein structure can correspond to such profound differences in organism. It is a sign of the foolishness into which an unreflective reductionism can lead us that we seriously argue from protein similarity to political similarity.

While they are more relevant to proteins than to politics, Darwin's writings have a great deal more in common with those other grand theorists of the nineteenth century, Marx and Freud, than with, say, Newton. Darwin's work is filled with ambiguities, contradictions, and theoretical revisions. The early Darwin of the *Beagle* in 1836 is neither the middle Darwin of the preliminary sketch of 1844 nor the mature Darwin of *The Origin* in 1859. Indeed, successive editions of *The Origin* contain important changes, and at one point Darwin seriously flirted with the inheritance of acquired characteristics, a notion that is logically fatal to his entire enterprise.

So, like the other Victorian radicals, Darwin has become the subject of a major historical industry. His letters, his diaries, his notebooks, his successive sketches, editions, and papers are the fossil bones to be used by the paleontologist of history in building a true picture of the beast. Unlike the remains of long-dead animals, however, the Darwinian fossil record is very unlikely to become fuller. Barring the discovery of a dusty bundle in some unlit corner of the Royal Society's attic, we have it all, and historians must be content to find the real Darwin by rereading and reinterpreting the same words. For me at least, the reconstruction has remained something of a cardboard cutout of a Great Man, eccentrically hypochondriacal, but indubitably a great man, unlike any practicing scientist I have ever known.

Barry Gale has changed all that. I do not know whether his thesis in *Evolution Without Evidence* that Darwin published *On the Origin of Species* without confidence in his evidence, and well before he intended to, is right or wrong. Certainly Gale has produced an abundance of quotations that support this view. As late as February 1858, Darwin wrote to Hooker, "I must come to some definite conclusion whether or not entirely to give up the ghost [of my theory]," and six months later wrote to Asa Gray at Harvard, "I cannot give you facts and I must write dogmatically, though I do not feel so on any point." But in a corpus as rich as the Darwin letters and notebooks, there are quotations to prove anything. What is appealing in Gale's work is a picture of a life in the social community of science that corresponds to our everyday experience of how careers are built.

64 Darwin returned from the voyage of the *Beagle* in 1836 to become a rising young star in geology. He was ambitious, courted success and successful men, and cared for their approval. He wrote in his autobiography (a genre not usually entered into by the self-effacing) that he wanted a "fair place among scientific men." When, after ten years, he had exhausted the career possibilities of geology, he turned his full attention to biology, including, among other questions, what was universally acknowledged to be *the* problem of the time. For twenty years he successfully exploited his relationships with the community of biologists to acquire information and specimens and to stake out a long-term claim on the species problem.

That is why it was to him that Alfred Russel Wallace wrote in 1858 with the news of his own independent invention of the theory of natural selection. Darwin was already a member of the British scientific establishment (he had received the Royal Medal of the Royal Society two years before), so it was to other establishment figures that he turned for tactical advice on how to save his scientific priority while saving his soul. He rather hoped the problem might go away since Wallace had not actually *said* he was hoping for publication, but Darwin's friends did not take the hint, and the solution agreed upon was a joint publication. So, it appears, Darwin was hustled into publication before he was really ready, for otherwise, as he put it, "all my originality, whatever it may amount to, will be smashed," and "it seems hard on me that I should be thus compelled to lose my priority of many years standing."[8]

It is not only ambiguity, contradiction, and long intellectual development that Darwin shared with other nineteenth-century revolutionaries. They are all dimly perceived through slogans. Survival of the fittest, like penis envy, is the opium of the people. To understand Darwinism simultaneously as a social phenomenon arising out of the remaking of the British social structure and as an extraordinary insight into the operation of natural forces requires considerable knowledge and subtlety of mind. To explain all that clearly, correctly, wittily, but without condescension to a lay public demands a high expository art. What one obviously needs for the job is to put together a physician-director-actor-comedian-TV star with the illustrator of *Swamp Comix*. It has been tried, and it works. While the illustrations are at times a bit swampier than the text demands, *Darwin for Beginners* is a superb introduction to a very tricky subject. It puts all the emphasis in the right place, is historically correct, scientifically impeccable, and contains as a postscript the best 250-word piece on reductionist social explanation yet written. Anyone who reads and understands Jonathan Miller's text will know a good deal more about Darwinism than most biologists and historians, while the pictures will be a constant reminder not to take the life of the mind more seriously than it deserves.

---

8. Letters to Charles Lyell, June 18 (the day he received Wallace's letter) and June 26, 1858.

What is the revolution that Darwin made? It was not the idea of evolution. Nor was it the invention of natural selection as an explanation. Although undoubtedly ingenious, and certainly a correct characterization of a great deal of evolution, it is, in the end, only a completion of the unfinished Cartesian revolution that demanded a mechanical model for all living processes. Nor was it even the variational model for a historical process in place of the usual transformational scheme. The invention of the variational model was indeed a considerable intellectual feat and represented a real epistemological break, for it changed the locus of historical action from the individual to the ensemble. Collectivities, the species, changed even though each individual within them was constant through its lifetime. What the variational model does is to convert one quality of variation, the static variation among objects in space, into another quality, the dynamic variation in time. As extraordinary as that insight may be, it can hardly be said to revolutionize, by itself, our way of seeing nature. It remains, again, a mechanism.

Darwin's real revolution consisted in the epistemological reorientation that had to occur before the variational mechanism could even be formulated. It was a change in the object of study from the average or modal properties of groups to the variation between individuals within them. That is, *variation itself* is the proper object of biological study, for it is the ground of biological being. Without it, there would have been no evolution and therefore no living biological world, for the earliest proto-life would have long since made the world uninhabitable for its own kind.

Before Darwin, the central issue for science was to discover the Platonic form that lay behind the imperfect reality, as Newton in the first book of the *Principia* treated ideal bodies moving in perfect voids, and only later considered the disturbing effects of friction and viscosity. Variation among organisms was thought to be ontologically distinct from the causes of their similarity, a similarity that we glimpse but dimly. If only we could eliminate the noisy confusion of the material objects themselves, the true relation would be seen. Darwin revolutionized our study of nature by taking the actual variation among actual things as central to the reality, not as an annoying and irrelevant disturbance to be wished away.

That revolution is not yet completed. Biology remains in many ways obdurately Platonic. Developmental biologists are so fascinated with how an egg turns into a chicken that they have ignored the critical fact that every egg turns into a different chicken and that each chicken's right side is different in an unpredictable way from its left. Neurobiologists want to know how the brain works, but they don't say whose brain. Presumably when you have seen one brain you have seen them all. Given the extraordinary complexity of connections in a brain, it is at least conceivable, if not likely, that two people may organize their memories of the same event differently, or, God forbid, differently on different days of the week. Even my cheap home computer reorganizes and moves its memory storage around as I add more input. Geneticists, who are supposed to know better, will sometimes talk about a gene's determining a particular shape, size, or behavior instead of reminding themselves that if genes determine anything, it is the pattern of variation of a developing organism in response to variation in the environment.

This error of geneticists is particularly ironic, because it was Gregor Mendel who, unknown to the rest of the scientific world, had, contemporaneously with Darwin, solved the other leading problem of biology by making variation his object of study. Mendel solved the problem of why offspring look like their parents by studying the pattern of differences between them. He discovered, as Darwin had, that similarity and variation are inextricably intertwined aspects of the same reality.

Epilogue

Had the article on Darwin's revolution been written in 1999 there would have been three differences from the text as it appeared sixteen years ago: two concerned with our knowledge of phenomena and one with the understanding of what was revolutionary in Darwin's view of evolution.

In *Evolution Now*, John Maynard Smith discussed the way in which non-Mendelian inheritance may be important in evolution. It was well known then that genes may move from one place on the chromosomes to another, carried on small stretches of DNA called transposable elements. A Nobel Prize was awarded in 1983 to Barbara McClintock for her discovery of this phenomenon. Since that time there has been an explosion not only in discoveries of new varieties of transposable elements but in our realization of how open the genome of a species is to the introduction of foreign DNA.

A simple transposable element consists of a short DNA sequence that matches target DNA sites distributed throughout the genes of their hosts, and a longer DNA sequence that codes for an enzyme that helps to insert or remove the transposable element from the host's genome. Thus, an element can pop into a host's genome in many different places and then pop out again. If it enters the host genome in the middle of a gene, that gene will now have an abnormal DNA sequence and so be mutated. Even if the element is later transposed again, it may leave a copy of itself behind or the excision may be imprecise so that the host's gene is still altered. Such transposable elements are extremely common and organisms' genomes are riddled with them. It is estimated that about 15 percent of the DNA in a fruit fly is composed of many copies of a variety of transposable elements.

A form of transposition that is of even greater significance for evolution than the movement of DNA within an organism is the horizontal transfer of genes between unrelated species by means of viruses or other cellular particles. Retroviruses are capable of copying DNA sequences of genes into their own molecular structure. The virus can then pass between quite unrelated organisms, for example by insect bites. After its introduction, the retrovirus can cause the new host to manufacture DNA copies of the gene information it has brought in and to incorporate that DNA into the host's own genome.

It used to be thought that new functions had to arise by mutations of the genes already possessed by a species and that the only way such mutations could spread was by the normal processes of reproduction. It is now clear that genetic material has moved during evolution from species to species by means of retroviruses and other transposable particles, although only a small number of cases have been studied. What is so extraordinary in its implications for evolution is that transposition can occur between forms of life that are quite different, between distantly related vertebrates, for example, or even between plants and bacteria.

Thus, baboons and cats, who have not had a common ancestor for about seventy million years, share a piece of DNA of currently unknown function that is not found in other mammals and that presumably was passed between them in relatively recent evolutionary time, perhaps by a mosquito. An enzyme, glucose phosphate isomerase, has passed from a close relative of phlox plants to bacteria, two forms that have not had a common ancestor in over a billion years. Thus, the assumption that species are on independent evolutionary pathways, once they have diverged from each other and can no longer interbreed, is incorrect. All life forms are in potential genetic contact and genetic exchanges between them are going on. The branches of the tree of life come together again.

A second discovery of the interconnections between life forms has been the growing evidence that the bits and pieces of the cells that constitute our bodies originated by the swallowing up of other organisms in the remote past. The cells of most organisms have in them so-called "organelles" within which certain important functions of the cell are carried out, such as the mitochondria that carry out energy metabolism and the chloroplasts of plants within which photosynthesis occurs. These organelles have their own DNA that codes for the enzymes that are operating within them. It is now apparent that such organelles are the descendants of what were once free-living bacteria or other very simple life forms, and that these either entered cells as parasites or were ingested by them and have long ago established a stable symbiotic relationship with their hosts. There is no evidence that such chimeras are still forming, but our understanding of the early history of life has been profoundly affected by these discoveries. The basic cellular architecture that is shared by yeasts and humans is a consequence of ancient fusions of independent life forms. Again, the evolutionary "tree of life" seems the wrong metaphor. Perhaps we should think of it as an elaborate bit of macramé.

The revolutionary step taken by Darwin is described in the original review as his invention of a variational scheme for evolution, as opposed to the older transformational one that characterized theories of cosmic evolution, cultural evolution, and biological evolution before his time. I have come to realize, however, that Darwin's variational theory, as revolutionary as it was, depended upon a prior and more radical epistemological break with the past. Darwin's theory of evolution was that variation among organisms arose from causes that were internal to the organisms and whose nature was independent of the demands of the external world. That is what is meant when we say that the mutations are "random." It is not that they are free from the ordinary processes of chemistry, but that their qualitative nature is at random with respect to how they will affect the organism in a particular environment. High temperature does not call forth mutations that specifically adapt the organism to live at high temperature. All sorts of mutations occur and it is only those that, by chance, enable the organism to survive better that will spread through the species. So the internal forces that give rise to variation are causally independent of the external forces that select them. The internal and the external, what we now think of as the gene and the environment, meet in the organism. This alienation of internal from external forces, of inside from outside, with the organism as their nexus, is fundamental to the Darwinian view. Indeed it is the origin of modern analytic biology.

Before the early nineteenth century internal and external forces were not seen as separated. Lamarck's scheme of evolution, a transformational one, assumed that the particular variations of form and function arose as a direct consequence of the needs of the organism to adapt to its outer world. Somehow the external forces molded the organism itself through its internal striving to adapt. This seamless connection between the inner and outer permeated views of nature. Not only were acquired characters inherited, but the very psychological state of a mother would influence its offspring. Jacob, it seems, took advantage of Laban's offer to let him keep all the multicolored sheep by holding a speckled stick before the eyes of ewes before they conceived. It is the alienation of the inside from the outside that has made modern analytic and reductionist biology possible. Certainly without that rupture we would have nothing of our modern understanding of the processes of evolution.

3

Chapter 3

Darwin, Mendel, and the Mind

"Darwin, Mendel & the Mind" was first published in The New York Review of Books *of October 10, 1985, as a review of* The Survival of Charles Darwin: A Biography of a Man and an Idea, *by Ronald W. Clark (Random House, 1984);* The Correspondence of Charles Darwin, Volume I: 1821–1836, *edited by Frederick Burkhardt and Sydney Smith (Cambridge University Press, 1985);* Past Masters: Mendel, *by Vítezslav Orel, translated by Stephen Finn (Oxford University Press, 1984);* Past Masters: Lamarck, *by L. J. Jordanova (Oxford University Press, 1984); and* Neuronal Man: The Biology of Mind, *by Jean-Pierre Changeux, translated by Dr. Laurence Garey (Pantheon, 1985).*

1.

THE CATALOG OF Harvard's Widener Library lists 184 books about Charles Darwin, his life and work (not counting 172 volumes of self-produced letters, autobiography, and scientific opera). On the subject of Gregor Mendel, there are only seventeen. The same disproportion is reflected in the books I have before me. Darwin is represented by a 702-page collection of letters all written before the age of twenty-seven, and a 449-page biography and subsequent history of the idea of evolution written by a professional biographer with no special expertise in the subject. When I contemplate yet another book about Darwin and Darwinism, I feel a bond of sympathy with the philistine Duke of Gloucester, whose reaction to a *second* volume of *The Decline and Fall* was, "Another damned, thick, square book! Always scribble, scribble, scribble, eh, Mr. Gibbon?" For Mendel on the other hand, the services of Vítezslav Orel, a great authority who has spent more than twenty-five years in historical research on the subject, have been obtained to produce a mere one hundred pages as part of a series of lives of the intellectual saints running from Aquinas to Wyclif.

As a population geneticist professionally concerned with Mendel's mechanism for the inheritance of variation and with Darwin's theory of evolution by selection of that variation, I have long found the vast disproportion in interest between the two to be paradoxical. While several explanations come to mind, none seems sufficient.

First, it might be argued that Darwin's popularity on the intellectual market is a classic case of consumer sovereignty. People are greatly concerned with the place of human beings in the universe, so the materialist theory of evolution continues to agitate and fascinate all concerned. After all, the first printing of *On the Origin of Species* was immediately sold out, and interest has hardly died out since, as evidenced by the legal and journalistic trials that occur at regular intervals in America. But the preoccupation of the literate middle classes and the fundamentalist masses with human uniqueness cannot explain the behavior of biologists, historians, and philosophers. While the hundredth anniversary, in 1959, of the publication of *On the Origin of Species* and the centenary, in 1982, of Darwin's death were the occasions for large numbers of international symposiums and their attendant publications, the 1965 centennial of Mendel's paper was lightly commemorated except in Czechoslovakia, and the centenary in 1984 of his death went completely unnoticed by the institutions of science. The *Journal of the History of Biology* would have to close its editorial offices if it were not constantly supplied with more and yet more about Darwin, but the *Folia Mendeliana*, almost a one-man industry of Dr. Orel's, appears only annually and is hard to find. In recent years, philosophers of science have abandoned physics for the richer and more complex domain of biology, which, God knows, needs their help, but they have almost all taken Darwinism as their focus of interest. The deep epistemological problems in heredity and development have been left largely to the philosophical naifs who practice the science.

Second, it might be claimed that Mendel's discovery was intrinsically less interesting, especially from a philosophical point of view, than Darwin's. The uncovering of the actual mechanism of heredity might be terribly important, but it is only a question of the mechanics, of particular gears and levers. But precisely the same can be said of Darwin. Although *On the Origin of Species* made evolution popular, Darwin certainly did not invent the idea. Indeed, a good case is made by L. J. Jordanova that if any biologist should be considered the father of evolutionary theory, it is Lamarck. Most French intellectuals have regarded the Anglo-Saxon infatuation with Darwin as a typical piece of chauvinism. Darwinism is, if anything, a particular mechanism for evolution. That mechanism is the differential rate of reproduction, under pressure from the environment, of different sorts of individuals within a population. Moreover, the success of Darwin's mechanical explanation of evolution depends critically on Mendel. Had heredity turned out to have a fundamentally different basis, Darwin's idea, ingenious though it was, would have been wrong.

The problem is that natural selection among variant types causes the population to lose variation as the superior type comes to characterize the species. That is, selection destroys the very population variation that is the basis for its operation. Evolution would then soon come to a stop if there were not some continued source of variation among individual organisms. If heredity takes place by a blending mechanism, either by the mixing of blood or other fluids, then any new variation that arises would be immediately diluted out by the process of mating and the production of intermediate hybrids. Darwin was acutely conscious of this problem of the loss of variation from blending inheritance and the constant need for new sources of variants. In later editions of the *Origin*, he allowed for the possibility that heritable variation could be directly induced by environmental action. That is, he took in Lamarck's view that acquired traits could be inherited, which is fatal to the whole Darwinian project of explaining evolution by a variational rather than a transformational mechanism. Mendelism saved the day.

The central core of Mendelism is the distinction between the *appearance* of an organism (the *phenotype*, in modern jargon), which may indeed be a blend of the characteristics of its parents, and the physical state of the factors inherited from each parent (the *genotype*), which remain physically discrete and unmixed. Just as Seurat's *Grande Jatte* gives the appearance of blended pigments from a close juxtaposition of small dots of pure color which are then visually fused by the physiology of the observer, so the physiology of development fuses, at the level of the whole organism, the pointillism of heredity.

Mendel's realization of this distinction came from his experimental crosses with garden peas. When he crossed two truebreeding varieties that differed markedly in some characteristic, say, flower color, the offspring were uniform in appearance, which is precisely what one would expect from a mixture of two varieties. In Mendel's case, there was the minor complication that the offspring all resembled one of the two parents rather than being intermediate between them, but this is the exception rather than the rule in most organisms. Thus, when Mendel crossed red-flowered and white-flowered garden peas, the offspring were all red-flowered. Had he worked with the sweet pea, *Lathyrus odoratus*, rather than the edible pea, *Pisum sativum*, the offspring would all have been pink. Whatever the color of the offspring flowers, the uniformity among individuals is precisely what one would predict from simple notions of the mixing of heredity. One would also predict that, if the uniform hybrids were crossed with each other, they would once again produce uniform offspring, and so on, without end.

But that is not what happened. When Mendel crossed these uniform hybrids with each other, he recovered in the next generation some plants with white flowers, like one of the two grandparents. From the reappearance of grandparental characteristics, apparently uncontaminated by their passage through the hybrids, and from the exact and repeatable ratios of types appearing among the offspring, Mendel constructed the two principles of heredity—principles that Darwinism needed to make it a workable theory. First, the factors that are passed from parent to offspring in heredity—what we now call genes—are particulate and maintain their individuality despite their interaction with other genes in the development of an organism. That is, the physical basis of heredity is discrete, like the elementary quantum of physics, rather than continuous.

Second, in the process of the formation of sperm and eggs in a hybrid organism, the genes that have been mixed together in that hybrid detach from each other and are parceled out to separate sperm and egg cells. That is the principle of segregation. Those two principles guarantee that if different variants in a population mate, even though their immediate offspring may be uniform and intermediate between the parents, in later generations the variation will reappear as a consequence of segregation. Thus new variation will not be submerged and diluted by the process of mating but will always be available for selection. Mendel's principle of segregation is the rock on which the theory of evolution by natural selection is built.

The real epistemological revolution wrought by Darwin was, in fact, identical with that created by Mendel. That identity can best be seen in the contrast with Lamarck, who was concerned with the problems both of evolution and of heredity. Jean-Baptiste

Lamarck was in many ways characteristic of the intellectual movement of the French Revolution. He was a deist, he accepted La Mettrie's *homme machine*, rejecting the soul, and assumed that material principles underlay all natural and human phenomena. He combined with his materialism, however, the eighteenth-century commitment to natural philosophy, to the principle that all of nature reflects a few general organizing principles. Jordanova discusses these principles and illustrates how they bore on the biological problems of evolution, taxonomy, and heredity, but, in her treatment, their origin remains mysterious. They seem a priori, or at least broad generalizations from a small base of observations. Some of the organizing principles that Lamarck espoused, such as the effect of use and disuse of organs, certainly hold across a reasonable domain of phenomena. Muscles do atrophy if they are not exercised and bones do grow larger and thicker at points where muscles are attached and produce tension. But brains do not atrophy with disuse and an important current theory of neurobiology (see below) actually maintains the opposite.

Other of Lamarck's principles, like the inheritance of acquired characteristics, were simply a priori or based on unexamined tradition. So giraffes' necks may indeed grow a bit longer if they stretch them to reach the tops of trees, but that change is not passed on to future generations. It took Darwin to see that giraffes that happened to be born with long necks were better able to survive than those without them. What Lamarck had in common with all natural philosophy was a typological view of phenomena, what Ernst Mayr has called "essentialism." In general, this meant that the ontological sources of similarity between things were seen as different from the ontological sources of differences. In particular, all members of a species were held to share unalterable properties that were intrinsic to the organisms, while differences between individual members were accidental consequences of environmental modification and were subordinate to the constant features. The problem of understanding the similarities was seen as fundamentally separate from the problem of the origin of superficial differences. It was the abolition of this distinction between the ontological sources of similarity and of difference that marked the epistemological break of Darwin and Mendel.

As I have argued in some detail in the previous essay, Darwin changed the object of study in evolution from the type of a species to the actual variation among individual organisms within the species. The motive power for a change in the average properties of the species was in the differences from the average displayed by the organisms themselves. Thus typical differences between species in space and in time arise by the accumulation of differences that were already present as variation *within* a species at any one place and time. But precisely the same contrast of similarities and differences permeated the study of heredity. Before Mendel, all studies of inheritance took heredity, that is, the passage of similarity between parents and offspring, to be different from, and antithetical to, the phenomenon of variation between individuals. The object of study was not the individual organism and its variant properties but the average or collective description of groups of progeny. For the predecessors of Mendel, the appearance of white-flowered plants among the progeny from the cross of the red-flowered plants implied a source of differences between organisms that was itself different from and obscured the action of the forces of heredity. For some, it was evidence of environmental malfunction. For others, it was the consequence of a poorly specified "force of atavism." It remained for Mendel to use the very occurrence of different types among the offspring of a single cross as the key to the laws of inheritance.

It is sometimes said that Mendel's unique contribution came because he was an experimentalist, or that he worked with favorable material, or that he had the sense to count the numbers of different types that came from his experimental crosses. But none of these is the critical element. Alexander Seton and John Goss in 1822 and Thomas Knight in 1823 had already observed segregation of green and pale seeds in second generations of pea crosses, Mendel's very material. Louis Vilmorin in 1856 counted the results from individual crosses and even reported three-to-one ratios of the two original parental types in the progeny of hybrids, the ratios upon which Mendel built his theory of the segregation of particulate factors in the formation of sex cells. Darwin observed "Mendelian" ratios in snapdragons but drew no conclusions. Even he did not realize that the variation among sister plants from the same parents was the proper object of study in heredity.

What Mendel understood, and what was not realized again for thirty-five years after his paper was made public, was that heredity and variation are two aspects of the same phenomenon and that only by a study of the actual variation among members of the same generation can we understand the passage of similarity across generations. This synthesis of the antithetical properties of heredity and variation is a Hegelian's dream and represents as difficult and subtle an insight into nature as any in the history of science.

A third reason for the vastly greater concentration on Darwin is simply that there is just a lot more information about him than there is about Mendel, so, of course, there is that much more grist for the academic mill. Darwin gave us an autobiography composed after he was already a very famous person, while Mendel produced only a curriculum vitae meant to support his application for a secondary teacher's certificate. There are hundreds of Darwin's letters, both personal and scientific, to scores of different recipients, including leading scientific figures, and he spent a couple of hours every day on his correspondence. Mendel is represented only by ten letters to the botanist Karl Nägeli, and a handful to his mother, sister, brother-in-law, and nephew, all of which were already available to Mendel's biographer, Hugo Iltis, in 1924.[1] While Darwin's sketches and notebooks remain as a rich

---

1. Hugo Iltis, *Gregor Mendel: Leben, Werk und Wirkung* (Berlin: J. Springer, 1924). In English as *Life of Mendel* (Norton, 1932).

source for historians, nearly all of Mendel's notes and papers were burned at his death in 1884, leaving us only a few reports to scientific and administrative committees on which he served, and the manuscript of two metaphysical poems written by the future priest when he was a schoolboy. In the absence of materials, what is there for a historian to do? A good deal, in fact, once the Great Man theory of history has been firmly put aside.

Although it has been a long while since political historians have abandoned the Suetonian ideal of history as biography, intellectual history has continued to concentrate on the individual genius as its proper focus. Did Darwin come to the idea of differential survival and reproductive success of units of different adaptive efficiency out of his head, in a true epistemological break? Or did he come to it for reasons external to scientific reasoning—for example, that his income was largely derived from stocks (mainly railroad shares) which he actively traded and whose rise and fall he followed daily, and with considerable care, in the newspaper?[2]

This struggle between internalist and externalist schools of historiography has not really changed the emphasis on individual genius because the argument has been about the *sources* of influence on the individual mind rather than on the central structure of causation. Even if external factors were predominant, we still call the modern theory of evolution "Darwinism," not "share marketism." Whatever the source of

2. I am indebted to Diane Paul for uncovering this fact for me.

influences, historians do not see the great mind of intellectual history as a mere passive nexus of external forces, but as the critical and central element in invention. Even the self-consciously historicist Marxist J. D. Bernal made individual scientists the leading actors in his recounting of the history of science.[3] Had Darwin not recovered from his attack of scarlet fever at the age of nine, would we be deprived of our understanding of natural selection? Well, not quite, since there was Alfred Wallace, there was Edward Blyth, and perhaps others that we know not of.

The standard view of Mendel makes him even more remarkable than Darwin. The son of a Moravian peasant freeholder, Mendel spent his entire life, except for a brief period of study in Vienna, in the provinces. Olomouc and Brno were not London and Cambridge. The Augustinian monastery where he lived and worked, and the technical high school in which he taught physics, housed a number of intellectuals, including a former professor of mathematics and physics at Lemberg (now Lvov) who had been the victim of the McCarthyism of the 1850s. But none was such an eminent scientist as Charles Lyell or Joseph Hooker, nor was there the great mass of active scientific workers in the mainstream of natural history like those with whom Darwin was in constant communication. The only scientist of note with whom Mendel corresponded was Nägeli, a contact he made only *after* he had completed his

3. J. D. Bernal, *Science in History* (London: Watts, 1954).

famous research on peas. Indeed, the lateness of his
exchange with Nägeli was Mendel's great luck. The
eminent professor induced the amateur to take up a
line of work with the hawkweed that cost Mendel five
years of frustration, strained his weakening eyesight,
and was bound to fail because of a peculiarity of the
sex life of that species of which Nägeli and Mendel
were unaware.

Mendel entered the monastery at Brno in 1843 at
the age of twenty-one because he was hard up and
saw no other way in which to complete his education
and to become a teacher. The standard picture of the
Königinkloster in Mendel's time is that of a sheltered
congregation of amateur intellectuals where the monk,
"alone, or in converse with the sage and tranquil
fathers...roamed through the monastery garden."[4]
Fortunately for Mendel, the monastery was presided
over by the Prelate Napp, "a man of large views, [who]
was delighted that the institution under his care should
become an intellectual center" where "almost all the
inmates...were engaged in independent activities,
either scientific or artistic."[5] On reading, as a young
student, Iltis's description I was tempted to enter an
order myself. A lack of space and the prelate's benign
multiplicity of interests, it seems, forced Mendel to
carry out his experiments on peas in a cramped garden
plot assigned him by the abbot next to the monastery
wall. Only when he himself became prelate in 1868

4. Iltis, *Life of Mendel*, p. 51.

5. Iltis, *Life of Mendel*, p. 49.

could Mendel enlarge his experimental domain. Thus we are led to a picture of a genteel intellectual atmosphere, encouraged, but within moderate bounds, by a "large-minded" abbot, the perfect bed for the germination and growth to maturity of Mendel's intellectual embryo.

In large part through the studies by Orel and others associated with the *Mendelianum* in Brno, this picture has changed drastically in the last twenty-five years. Moravia at the beginning of the nineteenth century was a rich agricultural region, especially important in sheep, fruit trees, and wine production. The revolutionary belief in the power of science to transform nature and promote economic and social progress was put into early practice in Moravia, largely at the instigation of Christian André, economic adviser to Count Salem. Together they founded the Agricultural Society, which pursued research in sheep breeding, pomiculture, and viniculture. A Pomological Association was begun in Brno as a branch of the Agricultural Society and, by the time Napp became prelate of the *Königinkloster* in 1825, breeding research in the region was an important activity.

Almost immediately, Napp established a fruit tree nursery in the monastery grounds and he wrote a manual on growing improved varieties. By 1830, the monastery was described as a "research establishment" for vine breeding. Napp was president of the Pomological Association. He successfully fought off an attempt to close down the monastery and replace its population with a contemplative order that would pay proper attention to the kingdom of God rather than the domain of Ceres. Napp recruited Mendel into the monastery by asking the professor of *physics* at the Olmütz (now Olomouc) Philosophical Institute to recommend one of his students to be a novice, and later sent his recruit to Vienna for further study in physics.

Far from grudgingly restricting Mendel to a narrow garden plot for the pursuit of his intellectual hobby, Napp built a greenhouse in 1855 in which Mendel and two research assistants from among the monks also worked. Plans for a much grander glass conservatory, never built, have survived. The monastery had a library of 20,000 volumes and an immense herbarium. Mendel was a founder of the Brünn (Brno) Society of Natural Science, to which he gave his famous paper on the laws of inheritance in 1865. When he succeeded Napp as prelate in 1868, Mendel continued the research work of the monastery. He had extensive and carefully designed experimental beehives built and was a founder of the Apiculture Society. He was a founding member of the Austrian Meteorological Society and spent many years in active weather observations which he regularly communicated. We recognize in Mendel the nineteenth-century version of the professional research scientist who, at the same time, as department chairman is in constant conflict with higher authorities on questions of budget and recruitment.

Thus the proper domain of historiography of Mendelism includes all of the scientific Enlightenment in Moravia, the structure and politics of local and regional scientific societies, and an understanding of the role of monasteries and of the Church as the controller of such institutions as the Brünn Philosophical Institute. As biography, historiography must be as concerned with the life and works of André, of Napp, of Franz Klácel, Mendel's utopian socialist fellow monk, a Hegelian who ended his days agitating in the Czech community of Illinois, as it is with the life and mind of Mendel himself. The difference between this community of institutions and personalities and the one in which Darwin was embedded was one of scale, wealth, and world influence. Britain, as the leading industrial and imperial power of the nineteenth century, also had a large positive balance of intellectual trade. Moravia was a net importer of ideas and, although the Brünn Society of Natural Science exchanged its publication with 130 other learned societies, it was to the *Proceedings of the Linnean Society* of London that one looked for the latest breakthrough. Modern historians of science seem still to be dazzled by Victorian values.

2.

Despite the growing dominance of externalist views of intellectual history, we remain preoccupied with the biographies of individual intellectuals. This attention to the lives and loves of great creators is, in part, rational. It is, after all, people and not societies who think. Mental images, concepts, trial solutions to problems, abstract orderings of the world are the proximate result of physiological processes that go on inside particular human beings. At the same time, of course, the social and natural world in which those beings are embedded are conditioned by and condition those individual thoughts. The formation of an idea is the individual transformation of a social condition by a material organism that is itself the product of individual and social conditions. The material basis of that transformation is the most difficult and most seductive problem in biology.

The problem for biologists trying to understand the human central nervous function is that they cannot, like the cardiac physiologist, proceed to the problems at hand without bothering their heads about abstruse philosophical issues. Any sensible study of integrated nervous function must come immediately to the mind–body problem that has been the perpetual agony of epistemology.

There is no unique way to describe or study a natural object. We begin always with a problem that sets the conditions of our description. A description of the heart and blood vessels that ignores their function in the circulation of the blood would be possible but quite beside the point. Unlike Aristotle, we do not believe that the brain's function is to cool the blood. Thoughts, feelings, perceptions are some kind of reflection of integrated central nervous activity, so whatever our theory of the brain, it must be so constructed that it helps us to think about thinking. Thus we cannot avoid, from the very beginning, the problem of the relationship between three pounds of flesh and the quality of mercy.

The reaction of neurobiologists to a confrontation with the mind–body problem has been largely to find one or another escape route. The molecular neurophysiologists can avoid the issue by concerning themselves entirely with the microscopic anatomy, chemistry, and physics of nerve conduction. What is a nerve impulse and how does it get from here to there? The elegant solution to this problem, which now seems more or less entirely in hand, depends only on accepting Descartes's machine model of the body without giving the slightest consideration to his dualism of body and mind. That is, one restricts one's questions to the domain where materialism is unchallenged. At the other extreme, we can, like Sir John Eccles,[6] give up materialism, accepting Descartes's *bête machine* but rejecting La Mettrie's *homme machine*, thus giving the mind an independent existence deriving from some nonmaterial source. It is a rare modern biologist, however, who is willing to make his or her peace with the soul. Many, if not most, subscribe to epiphenomenalism, a kind of backdoor dualism that gives primacy to the physical state of the brain as the *cause* of the mental which is, in itself, not causally efficacious. The steam engine and its whistle are the favored metaphor of this school.

The problem with epiphenomenalism is that it cannot cope with the evident effect of mental states on physical objects. Lauren Bacall's invitation to Humphrey

6. K. R. Popper and J. C. Eccles, *The Self and Its Brain* (London: Sprenger, 1977).

Bogart—"If you want anything, all you have to do is whistle"—would not work on a steam engine. Finally, one can deny the existence of the problem altogether by denying mind and claiming with J. B. Watson and B. F. Skinner that the concept of mind is hopelessly metaphysical and that only behavior exists. Precisely how the truth tables of Whitehead and Russell are to be described as behavior is not clear, unless one allows that they represent *mental* behavior, in which case we are back in the soup.

The only coherent materialist position seems to be that the mental and the neural are simply two aspects of the same material physical state. Mind neither causes a physical state of the brain nor is caused by it, since cause and effect do not apply to two aspects of the same state. This view conceives the program of neural research to be one of establishing the mapping of physical and mental onto each other. What configuration of neuronal connections and circulating currents corresponds to my mental image of Gregor Mendel? How did it get that way? What features in common does it have with your image of Mendel or, for that matter, with my image of Mendel last week?

While such a dual-aspect view of the mental and the physical has been expounded before,[7] it has not been so coherently and lucidly related to the newest facts of

7. See S. P. R. Rose, *The Conscious Brain* (Middlesex: Harmondsworth, 1976), and Chapter 10 in R. C. Lewontin, S. P. R. Rose, and L. J. Kamin, *Not in Our Genes* (Pantheon, 1984).

neuroanatomy and physiology as in the remarkable book of Jean-Pierre Changeux, *Neuronal Man*. Reading Changeux's book I had the sensation of watching him come toward me across a boggy moor, sure that at any moment he would step off the firm path and be swallowed, like his predecessors, in one or another quaking mire. Suddenly, almost to my envious disappointment, there he was beside me, smiling, never having made a false step. He begins with a description of the facts of neuroanatomy, the parts of the brain, and what is connected to what. As I read his description of the columns of brain tissue within which all the cells respond to stimulation of the same tissue, say skin, or bone joints, or to the left or right eye, I thought, "Now he will try to tell me that the brain is divided neatly into slabs, one for each function."

But instead, Changeux points out that the slabs, or modules, differentiating skin from joint sensitivity run at right angles to, and are crossed by, those that correspond to right and left eyes. That is, the very same bit of tissue may then correspond to two quite different sensory phenomena, just as the same stretch of DNA can code for two quite different proteins, depending upon where the reading frame begins. "Well," I thought, "he's got out of that one, but, reductionist that he is, he will misstep next time." But he never does. The details of the physics and chemistry of nerve conduction are given only as a way of providing a firm material basis for further argument.

The most dangerous terrain is in the description of specific molecules that can be used to induce rage, pleasure, orgasm, pain when acting on particular regions of the brain. The temptation is to claim that there is a specific chemical label for each kind of behavior and sensation. But Changeux shows that such is not the case, that "there is no transmitter for anger or pleasure, any more than there is an isolated center for them." At every point, Changeux counterposes the evidence for specific localization of particular function with evidence for diffuse control and interaction of parts. His message is that organization is everything, but that the organization of the brain is not simple.

The heart of Changeux's argument about body and mind concerns mental objects. Using empirical evidence, he argues that perceptions that come from external stimuli, say the eye, are the flows of nerve impulses among well-defined groups of cells. A flow structure among nerve cells is formed when I perceive, say, a tree. If at some later time I form a mental image of that tree, what is happening is a similarly structured flow of nerve impulses in some other group of cells with a similar topology of connection. It is the similarity of that topology which allows me to identify the two mental images and to "see" the tree in my "mind's eye."

It is here that Changeux comes closest to a simple (although perhaps correct) physical reductionism. To say that two things are topologically similar, we may mean very directly that there are two physical structures whose three-dimensional pattern of nodes and connecting links is geometrically the same, even though they may look superficially different, rather like two alternative sets of house wires that will allow one to turn on the same set of lights with the same combination of switches. Changeux's topological similarity is of this literal physical kind.

We might, however, mean by topological similarity, *logical* similarity based on totally different wiring diagrams that are physically quite incompatible, in the sense that two very different computer programs designed for completely different kinds of computers can both carry out the same computational task. It is entirely possible that two mental images may be computationally the same without being topologically similar physically. Indeed, the computer metaphor for the brain has had a powerful influence on studies of the central nervous system, and the science of artificial intelligence that exploits this metaphor is a major mode of investigation. Yet the relation between Changeux's wiring-diagram picture and the computer metaphor is not that of simple alternatives between which a decision must be made.

There are no "computer models" of the brain, at least none that anyone would take seriously. That is, there are no hardware models of the brain in which there are biological equivalents of random access memory, central processing units, error-checking circuits, and cathode-ray tubes. There are, however, "computational" or "computer program" models of mental processes which are then set working in actual physical computers, but everyone agrees that these are so-called simulations of actual thought processes. The problem is that we cannot test or reject the hypothesis that thought processes are like computer programs because there is no hypothesis to test. Any set of propositions that can be verbalized or symbolized, any picture that we can draw, any process of choosing among alternatives, of classifying and recognizing, can *necessarily* be represented by a computer algorithm after the fact. Computer programs are a form of abstract thought and have no independent existence. They can always be made congruent with our mental processes as they come into our conscious minds. Thus the purpose of computer simulation of mental processes cannot be to find out how brain and mind are related, but to systematize our understanding of mind, and so provide conditions that must be met by our picture of the brain.

How the system of mind is then mapped onto physical structures is a quite different question, the question that Changeux claims to illuminate. If we are lucky, Changeux will be right, and logical topologies will be reflected in physical congruences. Unfortunately, past experience suggests that things will be more complicated. When the organization of genes of related function was first studied, it appeared that there was a one-to-one correspondence between their physical order along the chromosome and the order in which the enzymes specified by them worked in metabolism. It appeared also that, at the front end of each lineup of genes, there was a special controlling region that turned genes on and off in response to environmental demands. Changeux himself made a key contribution to the building up of this picture.

We now know that this simple physical linearity is not the general rule but that there are many different physical arrangements of related genes and many different forms of control of their function, including the simplest arrangement first discovered. The biologist is constantly confronted with a multiplicity of detailed mechanisms for particular functions, some of which are unbelievably simple, but others of which resemble the baroque creations of Rube Goldberg. As François Jacob pointed out, evolution is less a sophisticated engineer than a home-workshop tinkerer. It seems likely that the physical organization of mental processes will bear the stamp of this natural-historical *bricolage*.

A good deal of *Neuronal Man* is taken up with the problems of the development of the set of nerve connections that underlie our memories and perceptions. Changeux makes a convincing, if not absolutely clinching, case for the selectional theory now fashionable. Turning Lamarck's principle of use and disuse on its head, Changeux argues that an unstimulated nervous system makes very large numbers of random multiple connections which remain labile for various periods of time. Experience, i.e., stimulation of the nervous system, in this view, causes differential elimination of various of these multiple connections, leaving in place only those that form a coherent structure. In the absence of stimulation, random death of cells and rupture of connections will leave the pathways in a permanent state of disarray. That is why we can only learn to speak if we hear others speak, and why after a certain age such learning is impossible, as in the case of "wild children" who never acquire language. The analogy is to a photographic image in which certain silver grains are fixed by exposure to light, the remainder being washed out in the developing process. In fact, however, the development of the brain cannot be purely selectional, since new neuronal connections are being made all through life, and their formation is, in part, stimulated by sensory experience.

With the selectional theory of mental development, neurobiology recapitulates a historical tension between the two modes of explanation for the fit of organisms to their environments. For nineteenth-century biology, the question was how species in the process of their evolution come to fit the demands of the environments in which they find themselves. The transformational theory that we associate with Lamarck, the theory that environment leaves its direct impress on each organism, gave way to Darwin's variational theory of the differential survival and reproduction of already existing variation. In the twentieth century, this struggle of explanations was first replayed in the field of immunology.

When our bodies are assaulted by foreign substances —bacteria, ragweed pollen, flu viruses, dog hair, bee sting toxin, or a transplanted heart—we produce specially adapted protein molecules, antibodies, that attach themselves to the invaders and make them liable to destruction. By and large, this ability to form antibodies is a good thing, for without it we would soon die from infection, as do the sufferers from Acquired Immune Deficiency Syndrome. The puzzle for biologists has been how we can produce on demand the repertoire of thousands of differently shaped antibody molecules, each neatly fitted to the molecular shape of a different invading substance. One theory, the "instructional," held that the invading substance itself served as a kind of die stamp, impressing a complementary shape on a malleable generalized antibody molecule. The alternative, "selectional," theory was that, before being challenged, our bodies already possess an immense array of differently shaped antibody molecules, all in low numbers, and that invasion by the foreign substance results in a rapid increase in the rate of production of just that antibody that matches the intruder.

During the last twenty years, the selectionist view has won out, in large part because of discoveries in molecular biology that have clarified how such an immense variety of antibody molecules can be produced spontaneously. One of the major contributors to the molecular biological vindication of the selectional theory was Gerald Edelman of Rockefeller University, who turns up, together with Changeux, as a principal proponent of the selectional theory for mental development. If he is right, he will wind up shaking the hand of King Carl Gustaf for the second time. Neurobiology has long held to the Lamarckian instructional principle without great success. Darwinism may now serve it better.

Changeux's successful traverse of a tricky terrain leaves much of it, perhaps the most interesting part, unvisited. If mental images are ordered structures of electrical oscillations among nerve cells, how do we account for the passage of our *attention* from one mental shape to another? Changeux treats consciousness as if it were simply the opposite of unconsciousness. But the heart of the problem of mind and brain is the shift of consciousness by what appears to us to be a willful act. As I tire of writing, I think first of the impending visit of a friend, then I strain to hear which Scarlatti sonata my wife is practicing, and then I return again to think about the relation of ego and mental images. I have passed among three very different mental states all under the control of the willful "I." Some kind of information about all these states must all the while have been resident in my *brain*, but only one at a time was in my *mind*. What chooses among them? "I." The central problem remains for neurobiology: What is "I"?

Epilogue

The great successes of molecular biology in under-standing the machinery of the cell have left the impression that no problem is beyond the analytic power of biologists and that complete knowledge of the organism is just around the corner. That is certainly the claim made for the Human Genome Project.[8] Such optimism, however, can only be maintained if the mind–brain problem is ignored, as it has been by the great mass of biologists. P. B. Medawar's characterization of science as the "art of the soluble" captures the important truth that ambitious scientists are usually too perspicacious to devote their lives to really difficult problems where they have little hope of success. When faced with questions that they do not know how to answer—like "How does a single cell turn into a mouse?" or "How did the structure and activity of Beethoven's brain result in Opus 131?"—the only thing that natural scientists know how to do is to turn them into other questions that they do know how to answer. That is, scientists do what they already know how to do.

In the fifteen years since the first French publication of Jean-Pierre Changeux's book there has been no major advance in understanding the relationship between mind and brain. Certainly there has been no rush of the cleverest practitioners to apply themselves to the problem, largely because they do not know

---

8. See Chapter 5.

what the question really means. Biologists know what a brain is, but they are as confused as ever about mind, leaving it to the philosophers to sort out whether consciousness really exists as a phenomenon to be explained.[9] Their sympathies are with the tradition, represented by Daniel Dennett, that all talk of "consciousness" is a metaphysical delusion on which we should stop wasting time. There are brain states and there are the manifest behaviors of the organism, behaviors that can be measured and verified by external observers. After all, if Dennett is right then biologists know what is to be done and how to do it. Only in this light can we understand the direction of neurobiological and neuropsychological research in the last fifteen years.

First, there has been a major industry of brain-imaging research. In former days the only way to map mental functions onto brain regions was to electrically stimulate parts of the brain and observe the subject's reaction, or to observe the consequences of traumatic injuries that have destroyed or disabled part of the brain. Antonio Damasio, in his book *Descartes' Error*,[10] uses the evidence from damage to the prefrontal area to show that feeling and thinking are linked aspects

---

9. The philosophical issue of consciousness and its relation to a material basis in the body has been discussed in a number of articles by John Searle, including a contentious exchange between Searle and Daniel Dennett, who represents the view that consciousness does not really exist as a phenomenon. These essays are collected in Searle's *The Mystery of Consciousness* (New York Review Books, 1997).

10. Grossett/Putnam, 1994.

of normal mental function. The phenomenon of "blindsight," discussed by Lawrence Weiskrantz in *Blindsight: A Case Study and Implications*,[11] shows that people can correctly report information presented to them visually, even though a brain lesion affecting part of their visual field prevents them from being conscious that they have seen anything. Such studies continue to challenge our simple notions of mental function, but precisely because they depend upon gross abnormalities, there is always the possibility that they will not lead us to an understanding of normal function, although, of course, they must be explicable by any theory of normal processes. It is now possible, however, to study the localization of normal brain activity by taking advantage of the chemistry and gross physiology of the brain. Regions of the brain in which a high rate of electrical activity is taking place will have a high rate of molecular activity to create electrical charge differences within and between nerve cells, and an increased blood flow. These hot spots of activity can be detected by electromagnetic "cameras" that form images of the brain.

One asks, then, which regions of the brain light up when particular mental functions are being performed. For example, it has been discovered that bilingual persons use the same areas of the brain for both languages if they became bilingual before the age of five, but different areas for the two languages if they learned the second language later. A very large literature has now

---

11. Oxford University Press, 1986.

accumulated on brain imaging because it is easy to do. The question is whether it tells us what we want to know. It is not clear what I will have learned about the physical basis of mental function, or whether consciousness exists, as opposed to mere behavior, when I know which region of my brain is active as I compose a sentence.

Second, a great deal of attention has been paid to language as a mental function because language is accessible to the outside observer and therefore lends itself to construal as a behavior. The reciprocal metaphors of the brain as a computer and the programming system of a computer as a language lead easily to the model of human language as an "algorithmic" behavior, that is, language as a set of rules for logical computation.[12] This is the sort of mentation with which scientists and linguists who aspire to be scientists feel at home. There has then been a considerable speculative literature using linguistic theory, the known facts of neuroanatomy, and the principles of evolutionary

---

12. The disagreement over whether language is merely algorithmic behavior is central to the exchange between Searle and Dennett. Searle uses his famous "Chinese Room" argument. We suppose that a non-Chinese speaker is passed slips of paper with Chinese sentences on them that require some response. The appropriate responses are all listed in a table, and the inhabitant of the room simply looks them up and transcribes them on a second piece of paper that he passes out to a waiting Chinese. Searle's claim is that what is going on between the people outside the room is fundamentally different from what is going on inside the room, because the inhabitant of the room still doesn't understand Chinese. That is, the process involves no meaning for him, only syntax, whereas the Chinese understand what is being communicated. Of course, if you deny the separate existence of "meaning," then the Chinese Room doesn't cut any ice.

biology to push for one theory or another of the biology of language, but none of these can be said to have given a new direction to research.[13]

---

13. The most interesting and biologically detailed of these is Terrence Deacon, *The Symbolic Species: The Co-evolution of Language and the Brain* (Norton, 1997). An attempt to argue from first principles, but without much biology, for the evolution of a specific language faculty is given in Steven Pinker, *How the Mind Works* (Norton, 1997).

Chapter 4

THE SCIENCE OF
METAMORPHOSES

"The Science of Metamorphoses" was first published in The New York Review of Books *of April 27, 1989, as a review of* Controlling Life: Jacques Loeb and the Engineering Ideal in Biology, *by Philip J. Pauly (Oxford University Press, 1987), and* Topobiology: An Introduction to Molecular Embryology, *by Gerald M. Edelman (Basic Books, 1988).*

1.

THE HISTORY OF biology is the history of struggles over the difference between the animate and the inanimate. Natural philosophy, through the Renaissance, and folk wisdom, for a much longer time, saw the entire natural world as a single interconnected system in which radical transformations of qualities of both living and non-living things were entirely credible. It was not merely that one inanimate kind of substance could, by alchemical transformation, be made into another, or that a vain boy could become a flower, but that the inanimate and the animate were interchangeable. Men could be petrified and marble statues turned to warm flesh in the embrace of their admirers. Papal staves put forth leaves, while moldy cheese and rags bred forth mice.

Aristotle believed animals could come from mud and that the animate and inanimate graded imperceptibly into one another on the *scala naturae*. But even the ancients were ambivalent about the ease with which inanimate matter could make that imperceptible transition. Despite Lucretius' assurance that "even today many animals spring from the earth, formed from the sun's heat and rain," it was not regarded as an everyday occurrence. In the *Metamorphoses*, Ovid gives few examples like the case of Pygmalion's statue and the creation of man from clay by Prometheus.

The raising of the dead by Jesus was, after all, the evidence of special powers and the leafing out of the Pope's staff a sign of special grace. Moreover, the transition from Aristotle's view to our present belief that the living are separated from the dead by a one-way bridge was a long and problematical one. Already in the seventeenth century William Harvey had declared *ex ovo omnia*, but the idea of spontaneous generation—that life originated from nonliving matter—was given a boost when Anton van Leeuwenhoek, looking through his microscope, saw a multitude of tiny living particles swimming by.

The Whig history of science we learn in school tells us that by the end of the eighteenth century Lazzaro Spallanzani had nailed down the case against spontaneous generation by his *experimental* approach to what was purely metaphysical speculation by wicked Aristotelians. But, as a matter of fact, the very same experiments done by others got the opposite result, supporting spontaneous generation. The disproofs of spontaneous generation that we now regard as definitive, those of Pasteur showing that microorganisms reproduce, were carried out in response to a public solicitation by the French Academy of Sciences for someone to finally settle the issue—in 1860.

We should not imagine, like the Whig historians of science, that the struggle over spontaneous generation was a story of the triumph of materialism and empiricism over superstition and a priori natural philosophy. On the contrary, nineteenth-century materialists took sides *against* the biogenetic law, the rule of "all life from life." For if there were an unbridgeable gap between the nonliving and the living, how could we explain the primal origin of life except by the infusion of a vital spirit into clay by a Promethean God? Moreover, that vital spirit, distinct from the known material forces of the universe, must be lurking in all living organisms, impalpable and unmeasurable. Nothing could be more antimaterialist than the claim for the uniqueness of life.

The struggle over spontaneous generation embodies the contradiction that plagues biology even today. On the one hand, mechanist materialist biology assumes that living beings are simply another form of the motion of matter and that the reductionist tactic of tearing matter into smaller and smaller bits will reveal all there is to know about life. On the other, biologists have never been able to create the living from the nonliving, nor do they even know where to begin. The biogenetic law seems as unbreakable as ever.

In support of the multibillion-dollar project to determine the DNA sequence of the entire set of human genes, the eminent molecular biologist Walter Gilbert has claimed that when we know the entire human genome we will know what it is to be human. But that must mean that at the very minimum we will know what it is to be living flesh. Yet neither Gilbert nor any other molecular biologist that I know of has suggested that knowing what a human being is will enable us to make one. That is, biologists, while believing that living organisms are nothing more than a form of matter and its motions, also believe that there is some principle of organization of living matter that is shared with no other natural assemblage of atoms. The question biologists keep asking themselves is "Why is this matter different from all other matter?"

The distinction between knowing what things are made of and knowing how to create or manipulate them permeates science and yet is a source of confusion for reductionist biologists. The problem is that in the very operation of determining what things are made of we take them to bits, and in ways that destroy the very relations that may be of the essence. "We murder to dissect." Nor is this true only for whole organisms or cells. Despite a knowledge of the structure of protein molecules down to the very placement of their atoms in exact three-dimensional space, we do not have the faintest idea of what the rules are for folding them up into their natural form. That does not prevent us, however, from being able to fold them up correctly in some cases by blind empirical fiddling. In the view of most biologists the disjunction between knowing what and knowing how is only a reflection of temporary ignorance, and, anyway, the successful manipulation of the world is only secondary to the primary goal of understanding. But that view has not been universal and some have regarded the control of life to be the object of the enterprise, putting the question of understanding quite aside. The most famous of the bearers of this "engineering ideal" in biology, Jacques Loeb, is the subject of Philip Pauly's superb book, *Controlling Life*. Rarely does a scientific biography so clearly illumine deep and long-lasting ideological differences in the conduct of scientific work.[1]

Jacques Loeb was trained as a medical doctor at the University of Strasbourg and came to the US in 1891. When he moved to the University of California at Berkeley in 1902 after ten years at the University of Chicago, a page-one headline in Hearst's *San Francisco Examiner* announced that "Illustrious Biologist Joins Faculty of State University," accompanied by a four-column artist's sketch of Professor Loeb and his magnifying glass. Of course, Hearst was a California booster and November 1902 was during a rare slack period when the United States, having completed one of those Caribbean adventures so loved by Mr. Hearst, had not yet started on its next. Nevertheless, even a Nobel Prize winner these days can hardly count on more than a one-column feature on the day of the big news, and is unlikely to be noticed by the local stringer for the *Times* when he decides to trade in State Street for Union Square.

The reason for Loeb's popular and journalistic fame had been announced just three years earlier in the *Chicago Tribune*: "Science Nears the Secret of Life:

---

1. For a review of Pauly's book that emphasizes questions of power and mastery, see David Joravsky, "Off to a Bad Start," *The New York Review*, November 19, 1987, pp. 17–20.

Professor Jacques Loeb Develops Young Sea Urchin by Chemical Treatment—Discovery That Reproduction by This Means Is Possible a Long Step Towards Realizing the Dream of Biologists, 'to Create Life in a Test Tube.' "

Indeed, for many Loeb *had* created "life in a test tube." The modern antiabortion movement did not invent the idea that life begins at the mystical moment of fertilization. The passive and comatose egg is quickened into life by the active wriggling sperm, like Sleeping Beauty recalled to life by a princely embrace. Loeb's successful induction of embryonic development without the benefit of sperm—"artificial parthenogenesis"—seemed closely akin to spontaneous generation.

Indeed, a headline in the Boston *Herald* completed the connection: "Creation of Life. Startling Discovery of Prof. Loeb. Lower Animals Produced by Chemical Means. Process May Apply to Human Species. Immaculate Conception Explained." The intoxication of the press was extraordinary. The confusion between artificial parthenogenesis and spontaneous generation might be expected, but the conflation of the doctrine of the Immaculate Conception with the doctrine of the Virgin Birth seems inexcusable in a Boston newspaper.[2]

As Pauly so clearly shows, Loeb's triumph was neither the accidental consequence of a program devoted to broader ends nor a critical step in an analytical project designed to "understand" development and repro-

---

2. Although not, perhaps, in Loeb, a Jew by birth and an atheist by conviction, who made the same parallel.

duction. It was, rather, "a natural consequence of his conviction that biology was and should be an engineering science concerned with transforming the natural order." It was the coming together of the nineteenth-century ideological commitments to materialism, on the one hand, and an optimistic progressivism, on the other. The phenomenal world was material and only material, and through the workings of human intellect that material world could be manipulated for any desired end. It was not that all things lay within the possibility of human *understanding* but that they lay within the sphere of human *action*. Indeed, we owe to Loeb this extraordinary epistemological position, an extension of Ernst Mach's operationalist view that

> the proof of the explicability of any single life phenomenon is furnished as soon as it is successfully controlled unequivocally through physical or chemical means or is repeated *in all details* with nonliving materials. . . .
>
> We cannot allow any barrier to stand in the path of our complete control and thereby understanding of the life phenomena. I believe that anyone will reach the same view who considers the *control* of natural phenomena is the essential problem of scientific research.[3]

Moreover, such control of life was the object of the entire enterprise: "I believe that it can only help science

---

3. Jacques Loeb, *Die Umschau* 7 (1903), pp. 21, 25, quoted by Pauly.

if younger investigators realize that experimental abio-genesis is the goal of biology."[4]

Pauly notes the inevitable journalistic comparison of Loeb with Victor Frankenstein, but he makes nothing of the problem that the engineering ideal raises, the problem of the unintended consequences of pragma-tism that is the central theme of Mary Shelley's *Frank-enstein, or the Modern Prometheus*. Shelley and her husband were greatly interested in and greatly dis-turbed by the instrumentalism of the most eminent and influential English scientist of the early nineteenth cen-tury, Sir Humphry Davy. Davy's *Discourse, Introduc-tory to a Course of Lectures on Chemistry*, which Mary Shelley read just before beginning her own *Franken-stein*, was the inspiration for her fictional Professor Waldman, Frankenstein's teacher and model.

Davy's scientific work was a series of diverse re-searches in chemistry, biology, and practical physics that were often instigated by practical demands. He made discoveries in agricultural chemistry and invented the Davy lamp, which allowed miners safe illumination in gas-filled mines. He investigated the electricity of the torpedo fish and the composition of ancient coloring materials. He was the very model of a modern scientist general, solving the mysteries of nature for the benefit of human life. But his philosophical writings on chem-istry show that for him, as for Loeb, understanding was simply control; scholarship was secondary to artisanry.

---

4. Jacques Loeb, *The Dynamics of Living Matter* (Macmillan, 1906), p. 223.

It was this untheoretical pragmatism that was Frankenstein's error. In a blinding moment of truth he discovered a single secret of how to create life and then on "a dreary night of November" (November was the month of Loeb's announcement!), "I collected the *instruments of life* around me, that I might infuse a spark of being into the lifeless thing that lay at my feet" (emphasis added). While Shelley does not say, those "instruments of life" were surely the apparatus of galvanic electric stimulation with which Galvani's nephew, Aldini, in a public demonstration in 1803, had made the jaw of a hanged murderer quiver, his fist clench, and his eye open.[5] But the possession of this single trick is like the secret ingredient of an alchemical formula or the secret incantation of a sorcerer. It can call spirits from the vasty deep, but it cannot control them because there is no understanding of their true nature.

In the most revealing chapter for the intellectual historian, Pauly traces the pedigree of Loeb's empiricism into later generations. Although Loeb continued to work on parthenogenesis, adjusting the chemical condition to

---

5. For an incomparable discussion of the origins and ideological underpinnings of Shelley's *Frankenstein*, see Anne K. Mellor, *Mary Shelley* (Methuen, 1988), especially Chapters four, five, and six. Whether, as she and others suggest, Shelley got the idea of a "spark of life" from reading Ovid's verses on Prometheus' creation of man from clay seems to me open to question. If she depended on Dryden's ornate translation with its gratuitous "particles of Heavenly fire" and "Aethereal Energy," the suggestion is reasonable, but we know from Shelley's journals that she worked laboriously through some part of the *Metamorphoses* in Latin, and Ovid speaks only of *semina* (seeds) that may have been already present in the clay.

optimize the process, he also returned to his earlier work on animal behavior in a similar mode. While others tried to analyze behavior as being the result of intrinsic physiological processes, Loeb's instrumentalist view led him to emphasize the importance of the environment in eliciting responses from organisms. Just as in the case of reproduction, the goal of experimental science was to be the control of behavior by the appropriate external stimuli rather than a "metaphysical" program of analyzing the internal states of an organism.

John B. Watson, who founded behaviorism, was a disciple of Loeb's at Chicago. In his behaviorist manifesto of 1912, Watson declared that psychology's "theoretical goal is the prediction and control of behavior." And the network grew. Another disciple of Loeb's was W. J. Crozier, a physiologist at Harvard, and two of his students were Gregory Pincus, who erroneously claimed to have produced parthenogenesis in rabbits on his way to inventing the contraceptive pill, and B. F. Skinner, the inheritor and elaborator of Watson's behaviorism. Yet another admirer of Loeb's was H. J. Muller, who won the Nobel Prize for his discovery that mutations could be produced by x-rays. While Loeb wanted biology to create life, Muller's goal was to change it, to produce directed evolution both by controlling the process of mutation and by eugenic programs of controlled breeding. By general agreement a demigod of genetics, Muller never made a single contribution to the analysis of the underlying physiological, cellular, and biochemical mechanisms of inheritance. For him, as for Loeb, the control of life was the central issue.

Loeb's claim to be interested only in control, relegating analysis of internal mechanisms to the sphere of the "metaphysical," ran against the entire trend of the reductionist science of his day. He was under constant pressure to rationalize his work by the articulation of an analytical program, pressure to which he finally gave way. Between 1910 and 1918 Loeb gave up his radical Machian commitment to the control of life and integrated himself into the established epistemological order. "He no longer saw scientists as leaders in the transformation of the world, but as cloistral figures, removed from society, seeking pure knowledge." Prometheus was bound: more than bound, utterly transformed. The image of Loeb as the founder of the mechanistic concept of life comes down to us in the form of Max Gottlieb in Sinclair Lewis's *Arrowsmith*, the epitome of the pure scientist.

The struggle between control and analysis as the goal of biology continues to the present, albeit in a less disinterested form. The commitment to understanding, to knowledge as an end in itself, is deeply embedded in scholarly culture. It is a first cause of our collective scholarly existence. But as the easy problems of biology are solved, the costs of solving the remaining hard ones become greater and greater, and the strain on the credulity of those who pay has not been lessened by changing the tax brackets. It is simply impossible to justify the expenditure of a billion dollars on a project to put in sequence the complete DNA of a "typical" human being or corn plant on the grounds that it would be a lovely thing to behold. So are we assured that it is really all in the interest of curing cancer, relieving schizophrenia, and making groceries cheaper. What for Jacques Loeb was an honest epistemological position has become, for the postmodern Prometheus, a piece of direct-mail advertising.

2.

Living organisms are characterized by five properties: they reproduce, they evolve, they recognize themselves, they develop, and they feel. These properties have given rise to the five major problems in biological science. Three have turned out to be "easy" for mechanistic biology, and two are very hard. The problem of how organisms reproduce, how they pass on to their offspring the information that they are to be lions rather than lambs, is now largely solved thanks to Gregor Mendel and his molecular biological epigones. So, too, we know the global features of the evolutionary process, although local mysteries remain. The ways in which organisms recognize themselves as opposed to others are varied so that there is no single mechanism of self-recognition; but a major mode, the formation of antibodies in higher animals, is now well understood at the molecular and cellular level.

That leaves us with the two hard problems: What is going on inside my head as I write these words, and how, starting from a single fertilized egg in my mother's uterus, did I develop the brain, eyes, and fingers that make it all possible? The problem is not simply that we do not have single coherent stories to tell about these processes, but that we do not know how to produce well-framed questions of whose relevance we are sure. Instead we have faddish models that succeed each other at five- or ten-year intervals, driven largely by changes in available technology in other branches of science, rather than by any coherent intellectual program.

Even the "easy" problems were not, of course, all that easy, and great fame has accrued to those who like Mendel or Watson and Crick have made major contributions to the solution. Gerald Edelman became famous (at least within scientific circles) and won a Nobel Prize for his major contribution to solving the self-recognition problem at the molecular level. Like other molecular biologists, having succeeded in solving his "easy" problem at a fairly early age, he decided to try a "hard" one, and for the last dozen years he has worked on the formation of the central nervous system. As a consequence, working along lines related to those of Jean-Pierre Changeux[6] and others, he has produced a really new theory of how the thinking brain develops,[7] by analogy with Darwin's theory of evolution by natural selection. But the problem of the brain is a problem in development in a larger sense; it is a problem of how so immensely complex

6. See Chapter 3.

7. See Gerald Edelman, *Neural Darwinism: The Theory of Neuronal Group Selection* (Basic Books, 1987).

a cognitive system is created out of "rudiments of form and sense." And so, with an intellectual ambition unmatched since Sir Humphry Davy, Edelman slid into the other hard problem, embryonic development. *Topobiology* is his attempt to make a coherent story of embryonic development out of the vast research on the subject that has accumulated since the turn of the century.

The problem of embryonic development is often said to be that of the origin of heterogeneity from homogeneity. The problem arises at two levels. First, all sperm and eggs look pretty much alike, but even a child can tell a frog from a prince. That is, what is the origin of the immense difference between organisms starting with what appears to be pretty much the same stuff? Second, starting with a single rather dull-looking egg cell, what is "that strange eventful history" by which we acquire, only to lose in the end, our teeth, eyes, taste, everything? That is the problem of embryonic differentiation. With the hubris born of a surfeit of Nobel Prizes, the geneticists tell us that the answer is obvious, it's all in the genes. Frogs and princes have very different sorts of genes despite the superficial resemblance of their gametes, or reproductive cells, so we should expect the result of development to be very different. As for the development of differentiated tissues and organs within an individual organism, it is simply the expression of different members of our set of genes at different times and different places in the developing embryo.

For Edelman the difficulty of this facile nonexplanation is one of dimension. How, he asks, can one-dimensional information (the string of DNA) become manifest in a three-dimensional organism? Indeed, the problem seems to be even worse since an organism is really four-dimensional, changing in a patterned way with time. The time dimension is, in fact, expressed not in time units but in a sequential order of changes. So the standard description of the development of a frog is not in terms of hours and days but in terms of developmental stages with clear morphological markers (the one-cell, two-cell, four-cell stages; the first appearance of a nerve cord; and so on). This method of marking time is a consequence of another major feature of embryogeny. Although the absolute time rate of development is sensitive to external conditions like temperature, the major features of embryonic development follow each other in an invariant order.

Edelman's statement of the problem of development as one of dimension does not quite epitomize either the difficulty or his solution to it. In principle there is no difficulty at all in expressing instructions about three dimensions in a one-dimensional form. When I addressed the envelope in which this review was sent to "Robert Silvers, *New York Review of Books*, 250 West 57th Street," I did just that and in a very precise way, given Mr. Silvers's size relative to that of the rest of the universe. Moreover, had I added the instruction "To be opened immediately," I would have provided both temporal and functional information at the same time.

The problem is not one of dimension, but of size. The nucleus of a cell of the fruit fly *Drosophila*, the favorite organism of geneticists, has enough DNA to specify the structure of about five thousand different proteins, and about thirty times that much DNA is available to provide spatial and temporal instructions about when the production of proteins by those genes should be turned on and turned off. But this is simply too little, by many orders of magnitude, to tell every cell when it should divide, exactly where it should move next, and what cellular structures it should produce over the entire developmental history of the fly. One needs to imagine an instruction manual that will tell every New Yorker when to wake up, where to go, and what to do, hour by hour, day by day, for the next century. There is just not enough DNA to go around.

It must be, Edelman argues, that the present location of a cell and its present activity provide most of the information on what it is to do next. It is this contingency on position that makes biology into "topobiology." It is this contingency on position that explains why, in taking an organism to pieces, we lose its *organismal* property. Legs and arms have exactly the same kinds of skin, bone, muscle, hair, nail, and connective tissue in about the same proportions. And they are very similar in their overall dimensions and gross structure. Yet, luckily for us, whether they developed on our front or rear ends made the critical difference in their final form and function. Victor Frankenstein's real "instruments of life" are not electric generators and spark gaps, but microscopic compasses, rulers, and protractors.

The idea that the position of a bit of protoplasm relative to other bits provides critical developmental information is not a new one. On the contrary, most theories of embryonic development in the last seventy years have attempted to make something of the notion of positional information. Metaphors of "field" and "gradient" and spatial waves of chemical concentration have dominated embryology. The problem is that no one has been able to make these metaphors work, or to give them a material molecular basis. That is the task that Edelman sets for himself. His strategy is to push the notion of local positional information to its extreme by supposing that essentially all the action is at the level of small collectives of cells acting as a group on their immediate neighbors. There are no global gradients over the organism or large-scale fields in which cells are moving. Central planning has been replaced by local initiative in a kind of perestroika of the protoplasm.

In the spirit of the local autonomy of small collectives, Edelman divides the cellular processes of differentiation into two kinds. First there are cell population processes—the division, migration, and death of cells. Much of embryonic development, when given a bare-bones description, does indeed consist of differential rates of cell division and cell death and the movement of small clumps or sheets of cells from one place to another, accompanied by the folding and rolling of such sheets as they move. The remainder of development is *cytodifferentiation*, the qualitative change in the actual structure and function of individual cells.

Some cells, like those in our hair follicles, begin to pump out huge amounts of a particular protein; hence the demand for barbers. Others grow tiny hairlike appendages themselves, whose beating and waving keep microscopic dust from accumulating in our lungs.

During his work on the development of the central nervous system Edelman had his attention drawn to molecules that acted like glues between cells, and it is on these molecules that he builds the entire edifice of his general theory of development. These molecules, called CAMs (cell adhesion molecules) and SAMs (surface adhesion molecules), are turned on and off in cycles. By affecting the surface properties of cells they cause cells to aggregate; this is followed by the movements of sheets of cells across each other and along noncellular matrices of materials that are secreted by other cells. This is the origin of the large movements and foldings of tissue that give rise to the general shape of an organism and account for the location of tissue. In some unspecified way the surface interactions of cells with their neighbors then turn on special regulatory genes within cells, which, in turn, switch on and off the genes responsible for cytodifferentiation. At every stage it is the local interactions of cells and tissues that determine the further movement, division, and differentiation of cells in the locality, which lead to yet further new local interactions, and so on to adulthood.

But, you may object, how does an entire, integrated, functioning organism arise from this anarchy of local control? Where is the invisible hand? By asking that question, Edelman says, you reveal that you are still mired in nineteenth-century pre-Darwinian teleology. The invisible hand is natural selection. There is nothing *intrinsic* to the process of development that leads to an integrated functioning organism, any more than there is anything intrinsic to the process of mutation that leads to better-adapted organisms. Those developmental processes that lead to a nonfunctioning organism have been lost in evolution because the bearers of those faulty programs left no progeny. All that is left is the collection of local processes that give the *appearance* of overall coordination because they work. The invisible hand of development is the very one that the Scottish economists extended to Darwin. Although the molecular details of the process of development occupy the attention of Edelman, and are likely to be the center of attention for most of his biologically trained readers, it is his sweeping away of the teleological element lurking in most accounts of development that is most insightful and most radical. Although biologists constantly decry teleology, they have not been able to free themselves of it in their explanation either of development or of cognitive functions. By constructing a theory of development that is nothing but the collection of quasi-independent local events, filtered through natural selection, Edelman has offered reductionist biology its last chance of encompassing development in its epistemological program.

*T*opobiology is a hard book to read, even for a professional biologist. Part of the reason is that it is a hard subject with an immense phenomenology ranging from the anatomical to the molecular and with a relevant literature extending back into the mists of antiquity. But partly the subject has been obfuscated by Edelman's language. As I read, I thought of the famous sentence in Czech, "Put your finger down your throat," which is said not to have a single vowel. There are whole paragraphs in *Topobiology* without a monosyllable (including "and" and "the"). In a euphuistic frenzy of polysyllabism Edelman refers to chewing and swallowing as "mastication and deglutition." The author is certainly not an uncultured scientific idiot savant (as adolescents, he and I together learned to declaim Corneille, Racine, and lesser French classics at a school devoted to such manifestations of high culture). It may be that Edelman believes that large subjects demand large words. In some instances, at least, the imprecision of high-flown vocabulary seems designed to substitute for a precision of ideas. If so, it is too bad, because the importance of his project, and the opportunity it offers biologists to vindicate their faith in the analytic method where they have consistently failed before, would be more than enough justification for Edelman to use, repeatedly, those three little Anglo-Saxon words: "I don't know."

Epilogue

Like the problem of the biology of mind, the problem of how organisms develop from a single fertilized egg cell into that "cunning'st pattern of perfecting Nature" remains just outside the grasp of biology, because the question we know how to ask is not quite the question we ought to ask. Developmental genetics, carried out at the molecular level, has revealed an enormous amount about genes involved in development. Many of these genes have been identified. For example, we know the DNA sequences of a set of genes, the so-called *hox* genes, present in all animals, that are essential for head to-tail organization of the developing embryo. (It is not so clear what role they play in animals that do not have head and tail ends.)

Moreover, we know a great deal about when in development the information in the DNA sequences is transcribed by the cellular machinery and converted into the molecules that participate in developmental changes. The *transcription* process consists in copying the DNA sequence of a gene into multiple copies of a related molecule, RNA, whose sequence is complementary to the DNA, much as a photograph can be copied into a negative that has the same information as the original. The information in the multiple RNA copies will be used by the cell machinery for synthesizing particular proteins during development, a process called *translation*. We also know which parts of the developing embryo contain the molecules produced by transcribing and translating the different genes, and, most interesting of all, how the molecules produced from transcribing and translating one gene may physically interact with the DNA of other genes to influence whether the cells will transcribe those in turn. The program of developmental genetics and to a large extent all of developmental biology is to complete this picture of the network of signals that are passed from one gene to another during development. But when we have this complete picture of the signaling network we will not know why I have a head at the front end and legs at the back. It is no use to say that "head" genes were turned on in one place and "leg" genes in another. The first problem is to find out how the cells at each end "knew" where they were in the embryo and so knew which genes to transcribe. The second problem is how the proteins coded by those genes resulted in the particular patterns of cell division and movement and cell differentiation that made my head the shape it is, with hair instead of toenails.

Those are the questions posed by Edelman's *Topo-biology* and we seem no closer to answering them than we were ten years ago. There is no deep conceptual problem here, as there is in the case of the biology of mind. No one is in doubt that the shape and differentiation of organisms is the direct manifestation of the movements and agglomerations of molecules under the influence of yet other molecules. It is just that at present no one seems to know how to ask those questions in a form that leads to a research program, so they are swept under the rug.

Chapter 5

THE DREAM OF THE
HUMAN GENOME

"The Dream of the Human Genome" was first published in The New York Review of Books *of May 28, 1992, as a review of* The Code of Codes: Scientific and Social Issues in the Human Genome Project, *edited by Daniel J. Kevles and Leroy Hood (Harvard University Press, 1992)*; Mapping the Code: The Human Genome Project and the Choices of Modern Science, *by Joel Davis (Wiley, 1990)*; Mapping Our Genes: The Genome Project and the Future of Medicine, *by Lois Wingerson (Dutton, 1990)*; Genethics: The Ethics of Engineering Life, *by David Suzuki and Peter Knudtson (Harvard University Press, 1990)*; Mapping and Sequencing the Human Genome, *by the Committee on Mapping and Sequencing the Human Genome (National Academy Press, 1988)*; Genome: The Story of the Most Astonishing Scientific Adventure of Our Time—the Attempt to Map All the Genes in the Human Body, *by Jerry E. Bishop and Michael Waldholz (Simon and Schuster, 1990)*; Exons, Introns, and Talking Genes: The Science Behind the Human Genome Project, *by Christopher Wills (Basic Books, 1991)*; Dangerous Diagnostics: The Social Power of Biological Information, *by Dorothy Nelkin and Laurence Tancredi (Basic Books, 1989)*; *and* DNA Technology in Forensic Science, *by the Committee on* DNA *Technology in Forensic Science (National Academy Press, 1992)*.

1.

FETISH . . . AN INANIMATE object worshipped by savages on account of its supposed inherent magical powers, or as being animated by a spirit. (OED)

Scientists are public figures, and like other public figures with a sense of their own importance, they self-consciously compare themselves and their work to past monuments of culture and history. Modern biology, especially molecular biology, has undergone two such episodes of preening before the glass of history. The first, characteristic of a newly developing field that promises to solve important problems that have long resisted the methods of an older tradition, used the metaphor of revolution. Tocqueville observed that when the bourgeois monarchy was overthrown on February 24, 1848, the Deputies compared themselves consciously to the "Girondins" and the "Montagnards" of the National Convention of 1793.

> The men of the first Revolution were living in every mind, their deeds and words present to

every memory. All that I saw that day bore the visible impress of those recollections; it seemed to me throughout as though they were engaged in acting the French Revolution rather than continuing it.

The romance of being a revolutionary had infected scientists long before Thomas Kuhn made Scientific Revolution the shibboleth of progressive knowledge. Many of the founders of molecular biology began as physicists, steeped in the lore of the quantum mechanical revolution of the 1920s. The Rousseau of molecular biology was Erwin Schrödinger, the inventor of the quantum wave equation, whose *What Is Life?* was the ideological manifesto of the new biology. Molecular biology's Robespierre was Max Delbruck, a student of Schrödinger, who created a political apparatus called the Phage Group, which carried out the experimental program. A history of the Phage Group written by its early participants and rich in the consciousness of a revolutionary tradition was produced twenty-five years ago.[1]

The molecular biological revolution has not had its Thermidor, but on the contrary it has ascended to the state of an unchallenged orthodoxy. The self-image of its practitioners and the source of their metaphors have changed accordingly, to reflect their perception of tran-

---

1. *Phage and the Origins of Molecular Biology*, edited by J. Cairn, G. S. Stent, and J. D. Watson (Cold Spring Harbor Laboratory of Quantitative Biology, 1966).

scendent truth and unassailable power. Molecular biology is now a religion, and molecular biologists are its prophets. Scientists now speak of the "Central Dogma" of molecular biology, and Walter Gilbert's contribution to the collection *The Code of Codes* is entitled "A Vision of the Grail." In their preface, Daniel Kevles and Leroy Hood take the metaphor with straight faces and no quotation marks:

> The search for the biological grail has been going on since the turn of the century, but it has now entered its culminating phase with the recent creation of the human genome project, the ultimate goal of which is the acquisition of all the details of our genome.... It will transform our capacities to predict what we may become....
>
> Unquestionably, the connotations of power and fear associated with the holy grail accompany the genome project, its biological counterpart.... Undoubtedly, it will affect the way much of biology is pursued in the twenty-first century. Whatever the shape of that effect, the quest for the biological grail will, sooner or later, achieve its end, and we believe that it is not too early to begin thinking about how to control the power so as to diminish—better yet, abolish—the legitimate social and scientific fears.

It is a sure sign of their alienation from revealed religion that a scientific community with a high concentration of Eastern European Jews and atheists has chosen for its central metaphor the most mystery-laden object of medieval Christianity.

As there were legends of the *Saint Graal* of Perceval, Gawain, and Galahad, so there is a legend of the Grail of Gilbert. It seems that each cell of my body (and yours) contains in its nucleus two copies of a very long molecule called deoxyribonucleic acid (DNA). One of these copies came to me from my father and one from my mother, brought together in the union of sperm and egg. This very long molecule is differentiated along its length into segments of separate function called genes, and the set of all these genes is called, collectively, my genome.

What I am, the differences between me and other human beings, and the similarities among human beings that distinguish them from, say, chimpanzees, are determined by the exact chemical composition of the DNA making up my genes. In the words of a popular bard of the legend, genes "have created us body and mind."[2] So when we know exactly what the genes look like we will know what it is to be human, and we will also know why some of us read *The New York Review* while others cannot get beyond the *New York Post*. "Genetic variations in the genome, various combinations of different possible genes ... create the infinite variety that we see among individual members of a species," according to Joel Davis in *Mapping the Code*. Success or failure, health or disease, madness or sanity, our ability to take it or leave it alone—all are determined, or at the very least are strongly influenced, by our genes.

2. Richard Dawkins, *The Selfish Gene* (Oxford University Press, 1976), p. 21.

The substance of which the genes are made must have two properties. First, if the millions of cells of my body all contain copies of molecules that were originally present only once in the sperm and once in the egg with which my life began, and if, in turn, I have been able to pass copies to the millions of sperm cells that I have produced, then the DNA molecule must have the power of *self-reproduction*. Second, if the DNA of the genes is the efficient cause of my properties as a living being, of which I am the result, then DNA must have the power of *self-action*. That is, it must be an active molecule that imposes specific form on a previously undifferentiated fertilized egg, according to a scheme that is dictated by the internal structure of DNA itself.

Because this self-producing, self-acting molecule is the ground of our being, "precious DNA" must be guarded by a "magic shield" against the "hurricane of forces" that threaten it from the outside, according to Christopher Wills, by which he means the bombardment by the other chemically active molecules of the cell that may destroy the DNA. It is not idly that DNA is called the Grail. Like that mystic bowl, DNA is said to be regularly self-renewing, providing its possessors with sustenance "*sans serjant et sans seneschal*," and shielded by its own Knights from hostile forces.

How is it that a mere molecule can have the power of both self-reproduction and self-action, being the cause of itself and the cause of all the other things? DNA is composed of basic units, the *nucleotides*, of which there are four kinds, adenine, cystosine, guanine, and thymine

(A, C, G, and T) and these are strung one after another in a long linear sequence which makes a DNA molecule. So one bit of DNA might have the sequence of units . . . CAAATTGC . . . and another the sequence . . . TATCGCTA . . . and so on. A typical gene might consist of 10,000 basic units, and since there are four different possibilities for each position in the string, the number of different possible kinds of genes is a great deal larger than what is usually called "astronomically large." (It would be represented as a 1 followed by 6,020 zeros.) The DNA string is like a code with four different letters whose arrangements in messages thousands of letters long are of infinite variety. Only a small fraction of the possible messages can specify the form and content of a functioning organism, but that is still an astronomically large number.

The DNA messages specify the organism by specifying the makeup of the proteins of which organisms are made. A particular DNA sequence makes a particular protein according to a set of decoding rules and manufacturing processes that are well understood. Part of the DNA code determines exactly which protein will be made. A protein is a long string of basic units called amino acids, of which there are twenty different kinds. The DNA code is read in groups of three consecutive nucleotides, and to each of the triplets AAA, AAC, GCT, TAT, etc., there corresponds one of the amino acids. Since there are sixty-four possible triplets and only twenty amino acids, more than one triplet matches the same amino acid (the code is "redundant"). Another part of the DNA determines when in development and where in the organism the manufacture of a given protein will be "turned on" or "turned off." By turning genes on and off in the different parts of the developing organism at different times, the DNA "creates" the living being, "body and mind."

But how does the DNA recreate itself? By its own dual and self-complementary structure (as the blood of Christ is said to be renewed in the Grail by the dove of the Holy Ghost). The string of nucleic acids in DNA that carries the message of protein production is accompanied by another string helically entwined with it and bound to it in a chemical embrace. This DNA doppelgänger is matched nucleotide by nucleotide with the message strand in a complementary fashion. Each A in the message is matched by a T on the complementary strand, each C by a G, each G by a C, and each T by an A.

Reproduction of DNA is, ironically, an uncoupling of the mated strands, followed by a building up of a new complementary strand on each of the parental strings. So the self-reproduction of DNA is explained by its dual, complementary structure, and its creative power by its linear differentiation.

The problem with this story is that although it is correct in its detailed molecular description, it is wrong in what it claims to explain. First, DNA is not self-reproducing; second, it makes nothing; and third, organisms are not determined by it.

DNA is a dead molecule, among the most nonreactive, chemically inert molecules in the living world.

That is why it can be recovered in good enough shape to determine its sequence from mummies, from mastodons frozen tens of thousands of years ago, and even, under the right circumstances, from twenty-million-year-old fossil plants. The forensic use of DNA for linking alleged criminals with victims depends upon recovering undegraded molecules from scrapings of long-dried blood and skin. DNA has no power to reproduce itself. Rather it is produced out of elementary materials by a complex cellular machinery of proteins. While it is often said that DNA produces proteins, in fact proteins (enzymes) produce DNA. The newly manufactured DNA is certainly a *copy* of the old, and the dual structure of the DNA molecule provides a complementary template on which the copying process works. The process of copying a photograph includes the production of a complementary negative which is then printed, but we do not describe the Eastman Kodak factory as a place of self-reproduction.

No living molecule is self-reproducing. Only whole cells may contain all the necessary machinery for "self"-reproduction and even they, in the process of development, lose that capacity. Nor are entire organisms self-reproducing, as the skeptical reader will soon realize if he or she tries it. Yet even the sophisticated molecular biologist when describing the process of copying DNA lapses into the rhetoric of "self-reproduction." So Christopher Wills, in the process of a mechanical description of DNA synthesis, tells us that "DNA cannot make copies of itself *unassisted*" (emphasis added) and further that "for DNA to duplicate [itself], the double helix must be unwound into two separate chains. . . ." The reflexive verb formation creeps in unobserved.

Not only is DNA incapable of making copies of itself, aided or unaided, but it is incapable of "making" anything else. The linear sequence of nucleotides in DNA is used by the machinery of the cell to determine what sequence of amino acids is to be built into a protein, and to determine when and where the protein is to be made. But the proteins of the cell are made by other proteins, and without that protein-forming machinery *nothing* can be made. There is an appearance here of infinite regress (What makes the proteins that are necessary to make the protein?), but this appearance is an artifact of another error of vulgar biology, that it is only the genes that are passed from parent to offspring. In fact, an egg, before fertilization, contains a complete apparatus of production deposited there in the course of its cellular development. We inherit not only genes made of DNA but an intricate structure of cellular machinery made up of proteins.

It is the evangelical enthusiasm of the modern Grail Knights and the innocence of the journalistic acolytes whom they have catechized that have so fetishized DNA. There are, too, ideological predispositions that make themselves felt. The more accurate description of the role of DNA is that it bears information that is read by the cell machinery in the productive process. Subtly, DNA as information bearer is transmogrified successively into DNA as blueprint, as plan, as master plan, as master molecule. It is the transfer onto biology of the belief in the superiority of mental labor over the merely physical, of the planner and designer over the unskilled operative on the assembly line.

The practical outcome of the belief that what we want to know about human beings is contained in the sequence of their DNA is the Human Genome Project in the United States and, in its international analogue, the Human Genome Organization (HUGO), called by one molecular biologist "the UN for the human genome."

These projects are, in fact, administrative and financial organizations rather than research projects in the usual sense. They have been created over the last five years in response to an active lobbying effort, by scientists such as Walter Gilbert, James Watson, Charles Cantor, and Leroy Hood, aimed at capturing very large amounts of public funds and directing the flow of those funds into an immense cooperative research program.

The ultimate purpose of this program is to write down the complete ordered sequence of A's, T's, C's, and G's that make up all the genes in the human genome, a string of letters that will be three billion elements long. The first laborious technique for cutting up DNA nucleotide by nucleotide and identifying each nucleotide in order as it is broken off was invented fifteen years ago by Allan Maxam and Walter Gilbert, but since then the process has become mechanized. DNA can now be squirted into one end of a mechanical process and out the other end will emerge a four-color computer printout announcing "AGGACTT...." In the course of the genome project yet more efficient mechanical schemes will be invented and complex computer programs will be developed to catalog, store, compare, order, retrieve, and otherwise organize and reorganize the immensely long string of letters that will emerge from the machine. The work will be a collective enterprise of very large laboratories, "Genome Centers," that are to be specially funded for the purpose.

The project is to proceed in two stages. The first is so-called "physical mapping." The entire DNA of an organism is not one long unbroken string, but is divided up into a small number of units, each of which is contained in one of a set of microscopic bodies in the cell, the chromosomes. Human DNA is broken up into twenty-three different chromosomes, while fruit flies' DNA is contained in only four chromosomes. The mapping phase of the genome project will determine short stretches of DNA sequence spread out along each chromosome as positional landmarks, much as mile markers are placed along superhighways. These positional makers will be of great use in finding where in each chromosome particular genes may lie. In the second phase of the project, each laboratory will take a chromosome or a section of a chromosome and determine the complete ordered sequence of nucleotides in its DNA. It is after the second phase, when the genome project, *sensu strictu*, has ended, that the fun begins, for biological sense will have to be made, if possible, of the mind-numbing sequence of three billion A's, T's, C's, and G's. What will it tell us about health and disease, happiness and misery, the meaning of human existence?

The American project is run jointly by the National Institutes of Health (NIH) and the Department of Energy in a political compromise over who should have control over the hundreds of millions of dollars of public money that will be required. The project produces a glossy-paper newsletter distributed free, headed by a coat of arms showing a human body wrapped Laocoön-like in the serpent coils of DNA and surrounded by the motto "Engineering, Chemistry, Biology, Physics, Mathematics." The Genome Project is the nexus of all sciences. My latest copy of the newsletter advertises the free loan of a twenty-three-minute video on the project "intended for high school age and older," featuring, among others, several of the contributors to *The Code of Codes*, and a calendar of fifty "Genome Events."

None of the authors of the books under review seems to be in any doubt about the importance of the project to determine the complete DNA sequence of a human being. "The Most Astonishing Adventure of Our Time," say Jerry E. Bishop and Michael Waldholz; "The Future of Medicine," according to Lois Wingerson; "today's most important scientific undertaking," dictating "The Choices of Modern Science," Joel Davis declares in *Mapping the Code*.

Nor are these simply the enthusiasms of journalists. The molecular biologist Christopher Wills says that "the outstanding problems in human biology... will all be illuminated in a strong and steady light by the results of this undertaking"; the great panjandrum of

DNA himself, James Dewey Watson, explains, in his essay in the collection edited by Kevles and Hood, that he doesn't "want to miss out on learning how life works," and Walter Gilbert predicts that there will be "a change in our philosophical understanding of ourselves." Surely, "learning how life works" and "a change in our philosophical understanding of ourselves" must be worth a lot of time and money. Indeed, there are said to be those who have exchanged something a good deal more precious for that knowledge.

2.

Unfortunately, it takes more than DNA to make a living organism. I once heard one of the world's leaders in molecular biology say, in the opening address of a scientific congress, that if he had a large enough computer and the complete DNA sequence of an organism, he could compute the organism, by which he meant totally describe its anatomy, physiology, and behavior. But that is wrong. Even the organism does not compute itself from its DNA. A living organism at any moment in its life is the unique consequence of a developmental history that results from the interaction of and determination by internal and external forces. The external forces, what we usually think of as "environment," are themselves partly a consequence of the activities of the organism itself as it produces and consumes the conditions of its own existence. Organisms do not find the world in which they develop. They make it. Reciprocally, the internal forces are not autonomous, but act in response to the external. Part of the internal chemical machinery of a cell is only manufactured when external conditions demand it. For example, the enzyme that breaks down the sugar lactose to provide energy for bacterial growth is only manufactured by bacterial cells when they detect the presence of lactose in their environment.

Nor is "internal" identical with "genetic." Fruit flies have long hairs that serve as sensory organs, rather like a cat's whiskers. The number and placement of those hairs differ between the two sides of a fly (as they do between the left and right sides of a cat's muzzle), but not in any systematic way. Some flies have more hairs on the left, some more on the right. Moreover, the variation between sides of a fly is as great as the average variation from fly to fly. But the two sides of a fly have the same genes and have had the same environment during development. The variation between sides is a consequence of random cellular movements and chance molecular events within cells during development, so-called "developmental noise." It is this same developmental noise that accounts for the fact that identical twins have different fingerprints and that the fingerprints on our left and right hands are different. A desktop computer that was as sensitive to room temperature and as noisy in its internal circuitry as a developing organism could hardly be said to compute at all.

The scientists writing about the Genome Project explicitly reject an absolute genetic determinism, but they seem to be writing more to acknowledge theoretical possibilities than out of conviction. If we take seriously the proposition that the internal and external codetermine the organism, we cannot really believe that the sequence of the human genome is the grail that will reveal to us what it is to be human, that it will change our philosophical view of ourselves, that it will show how life works. It is only the social scientists and social critics, such as Kevles, who comes to the Genome Project from his important study of the continuity of eugenics with modern medical genetics; Dorothy Nelkin, both in her book with Laurence Tancredi and in her chapter in Kevles and Hood; and, most strikingly, Evelyn Fox Keller in her contribution to *The Code of Codes*, for whom the problem of the development of the organism is central.

Nelkin, Tancredi, and Keller suggest that the importance of the Human Genome Project lies less in what it may, in fact, reveal about biology, and whether it may in the end lead to a successful therapeutic program for one or another illness, than in its validation and reinforcement of biological determinism as an explanation of all social and individual variation. The medical model that begins, for example, with a genetic explanation of the extensive and irreversible degeneration of the central nervous system characteristic of Huntington's chorea may end with an explanation of human intelligence, of how much people drink, how intolerable they find the social condition of their lives, whom they choose as sexual partners, and whether they get sick on the job. A medical model of all human variation makes a medical model of normality, including social normality, and dictates a therapeutic or pre-emptive attack on deviance.

There are many human conditions that are clearly pathological and that can be said to have a unitary genetic cause. As far as is known, cystic fibrosis and Huntington's chorea occur in people carrying the relevant mutant gene irrespective of diet, occupation, social class, or education. Such disorders are rare: 1 in 2,300 births for cystic fibrosis, 1 in 3,000 for Duchenne's muscular dystrophy, 1 in 10,000 for Huntington's disease. A few other conditions occur in much higher frequency in some populations but are generally less severe in their effects and more sensitive to environmental conditions, as for example sickle cell anemia in West Africans and their descendants, who suffer severe effects only in conditions of physical stress. These disorders provide the model on which the program of medical genetics is built, and they provide the human-interest drama on which books like *Mapping Our Genes* and *Genome* are built. In reading them, I saw again those heroes of my youth, Edward G. Robinson curing syphilis in *Dr. Ehrlich's Magic Bullet*, and Paul Muni saving children from rabies in *The Story of Louis Pasteur*.

It is said that a wonder-rabbi of Chelm once saw, in a vision, the destruction by fire of the study house in Lublin, fifty miles away. This remarkable event greatly enhanced his fame as a wonderworker. Several days later a traveler from Lublin, arriving in Chelm, was greeted with expressions of sorrow and concern, not unmixed with a certain pride, by the disciples of the wonder-rabbi. "What are you talking about?" asked the traveler. "I left Lublin three days ago and the study house was standing as it always has. What kind of a wonder-rabbi is that?" "Well, well," one of the rabbi's disciples answered, "burned or not burned, it's only a detail. The wonder is he could see so far." We live still in an age of wonder-rabbis, whose sacred trigram is not the ineffable YWH but the ever-repeated DNA. Like the rabbi of Chelm, however, the prophets of DNA and their disciples are short on details.

According to the vision, we will locate on the human chromosomes all the defective genes that plague us, and then from the sequence of the DNA we will deduce the causal story of the disease and generate a therapy. Indeed, a great many defective genes have already been roughly mapped onto chromosomes and, with the use of molecular techniques, a few have been very closely located and, for even fewer, some DNA sequence information has been obtained. But causal stories are lacking and therapies do not yet exist; nor is it clear, when actual cases are considered, how therapies will flow from a knowledge of DNA sequences.

The gene whose mutant form leads to cystic fibrosis has been located, isolated, and sequenced. The protein encoded by the gene has been deduced. Unfortunately, it looks like a lot of other proteins that are a part of cell structure, so it is hard to know what to do next. The mutation leading to Tay-Sachs disease is even better understood because the enzyme specified by the gene has a quite specific and simple function, but no one has suggested a therapy. On the other hand, the gene mutation causing Huntington's disease has eluded exact location, and no biochemical or specific metabolic defect has been found for a disease that results in catastrophic degeneration of the central nervous system in every carrier of the defective gene.

A deep reason for the difficulty in devising causal information from DNA messages is that the same "words" have different meanings in different contexts and multiple functions in a given context, as in any complex language. No word in English has more powerful implications of action than "do." "Do it now!" Yet in most of its contexts "do" as in "I do not know" is periphrastic and has no meaning at all. While the periphrastic "do" has no *meaning*, it undoubtedly has a linguistic *function* as a placeholder and spacing element in the arrangement of a sentence. Otherwise, it would not have swept into general English usage in the sixteenth century from its Midlands dialect origin, replacing everywhere the older "I know not."

So elements in the genetic messages may have meaning, or they may be periphrastic. The code sequence GTAAGT is sometimes read by the cell as an instruction to insert the amino acids valine and serine in a protein, but sometimes it signals a place where the cell machinery is to cut up and edit the message; and sometimes it may be only a spacer, like the periphrastic "do," that keeps other parts of the message an appropriate distance from each other. Unfortunately, we do not know how the cell decides among the possible interpretations. In working out the interpretive rules, it would certainly help to have very large numbers of different gene sequences, and I sometimes suspect that the claimed significance of the genome sequencing project for human health is an elaborate cover story for an interest in the hermeneutics of biological scripture.

Of course, it can be said, as Gilbert and Watson do in their essays, that an understanding of how the DNA code works is the path by which human health will be reached. If one had to depend on understanding, however, we would all be much sicker than we are. Once, when the eminent Kant scholar Lewis Beck was traveling in Italy with his wife, she contracted a maddening rash. The specialist they consulted said it would take him three weeks to find out what was wrong with her. After repeated insistence by the Becks that they had to leave Italy within two days, the physician threw up his hands and said, "Oh, very well, Madam. I will give up my scientific principles. I will cure you today."

Certainly an understanding of human anatomy and physiology has led to a medical practice vastly more effective than it was in the eighteenth century. These advances, however, consist in greatly improved methods for examining the state of our insides, of remarkable advances in microplumbing, and of pragmatically determined ways of correcting chemical imbalances and of killing bacterial invaders. None of these depends on a deep knowledge of cellular processes or on any discoveries of molecular biology. Cancer is still treated by gross physical and chemical assaults on the offending tissue. Cardiovascular disease is treated by surgery whose anatomical bases go back to the nineteenth century, by diet, and by pragmatic drug treatment. Antibiotics were originally developed without the slightest notion of how they do their work. Diabetics continue to take insulin, as they have for sixty years, despite all the research on the cellular basis of pancreatic malfunction. Of course, intimate knowledge of the living cell and of basic molecular processes may be useful eventually, and we are promised over and over that results are just around the corner. But as Vivian Blaine so poignantly complained:

> *You promised me this*
> *You promised me that.*
> *You promised me everything*
>   *under the sun.*
>
>   • • •
>
> *I think of the time gone by*
> *And could honestly die.*

Not the least of the problems of turning sequence information into causal knowledge is the existence of large amounts of "polymorphism." While the talk in most of the books under review is of sequencing the human genome, every human genome differs from every other. The DNA I got from my mother differs by about one tenth of one percent, or about 3,000,000 nucleotides, from the DNA I got from my father, and I differ by about that much from any other human being. The final catalog of "the" human DNA sequence will be a mosaic of some hypothetical average person corresponding to no one. This polymorphism has several serious consequences. First, all of us carry one copy, inherited from one parent, of mutations that would result in genetic diseases if we had inherited two copies. No one is free of these, so the catalog of the standard human genome after it is compiled will contain, unknown to its producers, some fatally misspelled sequences which code for defective proteins or no protein at all. The only way to know whether the standard sequence is, by bad luck, the code of a defective gene is to sequence the same part of the genome from many different individuals. Such polymorphism studies are not part of the Human Genome Project and attempts to obtain money from the project for such studies have been rebuffed.

Second, even genetically "simple" diseases can be very heterogeneous in their origin. Sequencing studies of the gene that codes for a critical protein in blood clotting has shown that hemophiliacs differ from people whose blood clots normally by any one of 208 different DNA variations, all in the same gene. These differences occur in every part of the gene, including bits that are not supposed to affect the structure of the protein.

The problem of telling a coherent causal story, and of then designing a therapy based on knowledge of the DNA sequence in such a case, is that we do not know even in principle all of the functions of the different nucleotides in a gene, or how the specific context in which a nucleotide appears may affect the way in which the cell machinery interprets the DNA; nor do we have any but the most rudimentary understanding of how a whole functioning organism is put together from its protein bits and pieces. Third, because there is no single, standard, "normal" DNA sequence that we all share, observed sequence differences between sick and well people cannot, in themselves, reveal the genetic cause of a disorder. At the least, we would need the sequences of many sick and many well people to look for common differences between sick and well. But if many diseases are like hemophilia, common differences will not be found and we will remain mystified.

3.

The failure to turn knowledge into therapeutic power does not discourage the advocates of the Human Genome Project because their vision of therapy includes *gene* therapy. By techniques that are already available and need only technological development, it is possible to implant specific genes containing the correct gene sequence into individuals who carry a mutated sequence, and to induce the cell machinery of the recipient to use the implanted genes as its source of information. Indeed, the first case of human gene therapy for an immune disease—the treatment of a child who suffered from a rare disorder of the immune system—has already been announced and seems to have been a success. The supporters of the Genome Project agree that knowing the sequence of all human genes will make it possible to identify and isolate the DNA sequences for large numbers of human defects which could then be corrected by gene therapy. In this view, what is now an ad hoc attack on individual disorders can be turned into a routine therapeutic technique, treating every physical and psychic dislocation, since everything significant about human beings is specified by their genes.

However, gene implantation may affect not only the cells of our temporary bodies, our *somatic* cells, but the bodies of future generations through accidental changes in the *germ* cells of our reproductive organs. Even if it were our intention only to provide properly functioning genes to the immediate body of the sufferer, some of the implanted DNA might get into and transform future sperm and egg cells. Then future generations would also have undergone the therapy in absentia and any miscalculations of the effects of the implanted DNA would be wreaked on our descendants to the remotest time. So David Suzuki and Peter Knudtson make it one of their principles of "genethics" (they have self-consciously created ten of them) that

> while genetic manipulation of human somatic cells may lie in the realm of personal choice, tinkering with human germ cells does not. Germ-cell therapy, without the consent of all members of society, ought to be explicitly forbidden.

Their argument against gene therapy is a purely prudential one, resting on the imprecision of the technique and the possibility that a "bad" gene today might turn out to be useful some day. This seems a slim base for one of the Ten Commandments of biology, for, after all, the techniques may get a lot better and mistakes can always be corrected by another round of gene therapy. The vision of power offered to us by gene therapists makes gene transfer seem rather less permanent than a silicone implant or a tummy tuck. The bit of ethics in *Genethics* is, like a Unitarian sermon, nothing that any decent person could quarrel with. Most of the "genethic principles" turn out to be, in fact, prudential advice about why we should not screw around with our genes or those of other species. While most of their arguments are sketchy, Suzuki and Knudtson are the only authors among those under review who take seriously the problems presented by genetic diversity among individuals, and who attempt to give the reader enough understanding of the principles of population genetics to think about these problems.

Most death, disease, and suffering in rich countries do not arise from muscular dystrophy and Huntington's chorea, and, of course, the majority of the world's population is suffering from one consequence or another of malnutrition and overwork. For Americans, it is heart disease, cancer, and stroke that are the major killers, accounting for 70 percent of deaths, and about sixty million people suffer from chronic cardiovascular disease. Psychiatric suffering is harder to estimate, but before the psychiatric hospitals were emptied in the 1960s, there were 750,000 psychiatric inpatients. It is now generally accepted that some fraction of cancers arise on a background of genetic predisposition. That is, there are a number of genes known, the so-called *oncogenes*, that have information about normal cell division. Mutations in these genes result (in an unknown way) in making cell division less stable and more likely to occur at a pathologically high rate. Although a number of such genes have been located, their total number and the proportion of all cancers influenced by them is unknown.

In no sense of simple causation are mutations in these genes *the* cause of cancer, although they may be one of many predisposing conditions. Although a mutation leading to extremely elevated cholesterol levels is known, the great mass of cardiovascular disease has utterly defied genetic analysis. Even diabetes, which has long been known to run in families, has never been tied to genes and there is no better evidence for a genetic predisposition to it in 1992 than there was in 1952 when serious genetic studies began. No week passes without the announcement in the press of a "possible" genetic cause of some human ill which upon investigation "may eventually lead to a cure." No literate public is unassailed by the claims. The *Morgunbladid* of Reykjavik asks its readers rhetorically, "*Med allt í genunum?*" ("Is it all in the genes?") in a Sunday supplement.

The rage for genes reminds us of tulipomania and the South Sea Bubble in McKay's *Extraordinary Popular Delusions and the Madness of Crowds*. Claims for the definitive location of a gene for schizophrenia and manic-depressive syndrome using DNA markers have been followed repeatedly by retraction of the claims and contrary claims as a few more members of a family tree have been observed, or a different set of families examined. In one notorious case, a claimed gene for manic depression, for which there was strong statistical evidence, was nowhere to be found when two members of the same family group developed symptoms. The original claim and its retraction both were published in the international journal *Nature*, causing David Baltimore to cry out at a scientific meeting, "Setting myself up as an average reader of *Nature*, what am I to believe?" Nothing.

Some of the wonder-rabbis and their disciples see even beyond the major causes of death and disease. They have an image of social peace and order emerging from the DNA data bank at the National Institutes of Health. The editor of the most prestigious general American scientific journal, *Science*, an energetic publicist for large DNA-sequencing projects, in special issues of his journal filled with full-page multicolored advertisements from biotechnology equipment manufacturers, has visions of genes for alcoholism, unemployment, domestic and social violence, and drug addiction. What we had previously imagined to be messy moral, political, and economic issues turn out, after all, to be simply a matter of an occasional nucleotide substitution. While the notion that the war on drugs will be won by genetic engineering belongs to Cloud Cuckoo Land, it is a manifestation of a serious ideology that is continuous with the eugenics of an earlier time.

Daniel Kevles has quite persuasively argued in his earlier book on eugenics[3] that classical eugenics became transformed from a social program of general population improvement into a family program of providing genetic knowledge to individuals facing reproductive decisions. But the ideology of biological determinism on which eugenics was based has persisted and, as is made clear in Kevles's excellent short history of the Genome Project in *The Code of Codes*, eugenics in the social sense has been revivified. This has been in part a consequence of the mere existence of the Genome Project, with its accompanying public relations and the heavy public expenditure it will require. These alone validate its determinist *Weltanschauung*. The publishers declare the glory of DNA and the media showeth forth its handiwork.

4.

The nine books reviewed here are only a sample of what has been and what is to come. The cost of sequencing the human genome is estimated optimistically at 300 million dollars (ten cents a nucleotide for the three billion nucleotides of the entire genome), but if development costs are included it surely cannot be less than a half-billion in current dollars. Moreover the

---

3. Daniel J. Kevles, *In the Name of Eugenics: Genetics and the Uses of Human Heredity* (University of California Press, 1986).

genome project *sensu strictu* is only the beginning of wisdom. Yet more hundreds of millions must be spent on chasing down the elusive differences in DNA for each specific genetic disease, of which some 3,000 are now known, and some considerable fraction of that money will stick to entrepreneurial molecular geneticists. None of our authors has the bad taste to mention that many molecular geneticists of repute, including several of the essayists in *The Code of Codes*, are founders, directors, officers, and stockholders in commercial biotechnology firms, including the manufacturers of the supplies and equipment used in sequencing research. Not all authors have Norman Mailer's openness when they write advertisements for themselves.

It has been clear since the first discoveries in molecular biology that "genetic engineering," the creation to order of genetically altered organisms, has an immense possibility for producing private profit. If the genes that allow clover plants to manufacture their own fertilizer out of the nitrogen in the air could be transferred to maize or wheat, farmers would save great sums and the producers of the engineered seed would make a great deal of money. Genetically engineered bacteria grown in large fermenting vats can be made into living factories to produce rare and costly molecules for the treatment of viral diseases and cancer. A bacterium has already been produced that will eat raw petroleum, making oil spills biodegradable. As a consequence of these possibilities, molecular biologists have become entrepreneurs. Many have founded biotechnology firms funded by venture capitalists. Some have become very rich when a successful public offering of their stock has made them suddenly the holders of a lot of valuable paper. Others find themselves with large blocks of stock in international pharmaceutical companies who have bought out the biologist's mom-and-pop enterprise and acquired their expertise in the bargain.

No prominent molecular biologist of my acquaintance is without a financial stake in the biotechnology business. As a result, serious conflicts of interest have emerged in universities and in government service. In some cases graduate students working under entrepreneurial professors are restricted in their scientific interchanges, in case they may give away potential trade secrets. Research biologists have attempted, sometimes with success, to get special dispensations of space and other resources from their universities in exchange for a piece of the action. Biotechnology joins basketball as an important source of educational cash.

Public policy, too, reflects private interest. James Dewey Watson resigned in April as head of the NIH Human Genome Office as a result of pressure put on him by Bernardine Healey, director of the NIH. The immediate form of this pressure was an investigation by Healey of the financial holdings of Watson or his immediate family in various biotechnology firms. But nobody in the molecular biological community believes in the seriousness of such an investigation, because everyone including Dr. Healey knows that there are no financially disinterested candidates for Watson's job. What is really at issue is a disagreement about patenting the human genome. Patent law prohibits the patenting of anything that is "natural," so, for example, if a rare plant were discovered in the Amazon whose leaves could cure cancer, no one could patent it. But, it is argued, isolated genes are not natural, even though the organism from which they are taken may be. If human DNA sequences are to be the basis of future therapy, then the exclusive ownership of such DNA sequences would be money in the bank.

Dr. Healey wants the NIH to patent the human genome to prevent private entrepreneurs, and especially foreign capital, from controlling what has been created with American public funding. Watson, whose family is reported to have a financial stake in the British pharmaceutical firm Glaxo, has characterized Healey's plan as "sheer lunacy," on the grounds that it will slow down the acquisition of sequence information.[4] (Watson has denied any conflict of interest.) Sir Walter Bodmer, the director of the Imperial Cancer Research Fund and a major figure in the European genome organization, spoke the truth that we all know lies behind the hype of the Human Genome Project when he told *The Wall Street Journal* that "the issue [of ownership] is at the heart of everything we do."

The study of DNA is an industry with high visibility, a claim on the public purse, the legitimacy of a science, and the appeal that it will alleviate individual and social suffering. So its basic ontological claim, of the dominance of the Master Molecule over the body physical and the

4. See *The New York Times*, April 9, 1992; p. A26, *The Wall Street Journal*, April 17, 1992, p. 1; and *Nature*, April 9, 1992, p. 463.

body politic, becomes part of general consciousness. Evelyn Fox Keller's chapter in *The Code of Codes* brilliantly traces the percolation of this consciousness through the strata of the state, the universities, and the media, producing an unquestioned consensus that the model of cystic fibrosis is a model of the world. Daniel Koshland, the editor of *Science*, when asked why the Human Genome Project funds should not be given instead to the homeless, answered, "What these people don't realize is that the homeless are impaired. . . . Indeed, no group will benefit more from the application of human genetics."[5]

Beyond the building of a determinist ideology, the concentration of knowledge about DNA has direct practical social and political consequences, what Dorothy Nelkin and Laurence Tancredi call "The Social Power of Biological Information." Intellectuals in their self-flattering wish-fulfillment say that knowledge is power, but the truth is that knowledge further empowers only those who have or can acquire the power to use it. My possession of a Ph.D. in nuclear engineering and the complete plans of a nuclear power station would not reduce my electric bill by a penny. So with the information contained in DNA, there is no instance where knowledge of one's genes does not further concentrate the existing relations of power between individuals and between the individual and institutions.

---

5. Remarks made at the First Human Genome Conference in October 1989. Quoted by Keller in "Nature, Nurture, and the Human Genome Project," in *The Code of Codes*.

When a woman is told that the fetus she is carrying has a 50 percent chance of contracting cystic fibrosis, or for that matter that it will be a girl although her husband desperately wants a boy, she does not gain additional power just by having that knowledge, but is only forced by it to decide and to act within the confines of her relation to the state and her family. Will her husband agree to or demand an abortion, will the state pay for it, will her doctor perform it? The slogan "a woman's right to choose" is a slogan about conflicting relations of power, as Ruth Schwartz Cowan makes clear in her essay "Genetic Technology and Reproductive Choice: An Ethics for Autonomy" in *The Code of Codes*.

Increasingly, knowledge about the genome is becoming an element in the relation between individuals and institutions, generally adding to the power of institutions over individuals. The relations of individuals to the providers of health care, to the schools, to the courts, to employers are all affected by knowledge, or the demand for knowledge, about the state of one's DNA. In the essays by both Henry Greeley and Dorothy Nelkin in *The Code of Codes*, and in much greater detail and extension in *Dangerous Diagnostics*, the struggle over biological information is revealed. The demand by employers for diagnostic information about the DNA of prospective employees serves the firm in two ways. First, as providers of health insurance, either directly or through their payment of premiums to insurance companies, employers reduce their wage bill by hiring only workers with the best health prognoses.

Second, if there are workplace hazards to which employees may be in different degrees sensitive, the employer may refuse to employ those whom it judges to be sensitive. Not only does such employment exclusion reduce the potential costs of health insurance, but it shifts the responsibility of providing a safe and healthy workplace from the employer to the worker. It becomes the worker's responsibility to look for work that is not threatening. After all, the employer is helping the workers by providing a free test of susceptibilities and so allowing them to make more informed choices of the work they would like to do. Whether other work is available at all, or worse paid, or more dangerous in other ways, or only in a distant place, or extremely unpleasant and debilitating is simply part of the conditions of the labor market. So Koshland is right after all. Unemployment and homelessness do indeed reside in the genes.

Biological information has also become critical in the relation between individuals and the state, for DNA has the power to put a tongue in every wound. Criminal prosecutors have long hoped for a way to link accused persons to the scene of a crime when there are no fingerprints. By using DNA from a murder victim and comparing it with DNA from dried blood found on the person or property of the accused, or by comparing the accused's DNA with DNA from skin scrapings under the fingernails of a rape victim, prosecutors attempt to link criminal and crime. Because of the polymorphism of DNA from individual to individual, a definitive identification is, in principle, possible. But, in practice, only a bit of DNA can be used for identification, so there is some chance that the accused will match the DNA from the crime scene even though someone else is in fact guilty.

Moreover, the methods used are prone to error, and false matches (as well as false exclusions) can occur. For example, the FBI characterized the DNA of a sample of 225 FBI agents and then, on a retest of the same agents, found a large number of mismatches. Matching is almost always done at the request of the prosecutor, because tests are expensive and most defendants in assault cases are represented by a public defender or court-appointed lawyer. The companies who do the testing have a vested commercial interest in providing matches, and the FBI, which also does some testing, is an interested party.

Because different ethnic groups differ in the frequency of the various DNA patterns, there is also the problem of the appropriate reference group to whom the defendant is to be compared. The identity of that reference group depends in complex ways on the circumstances of the case. If a woman who is assaulted lives in Harlem near the borderline between black, Hispanic, and white neighborhoods at 110th Street, which of these populations or combination of them is appropriate for calculating the chance that a "random" person would match the DNA found at the scene of the crime? A paradigm case was tried last year in Franklin County, Vermont. DNA from blood stains found at the scene of a lethal assault matched the DNA of an accused man.

The prosecution compared the pattern with population samples of various racial groups, and claimed that the chance that a random person other than the accused would have such a pattern was astronomically low.

Franklin County, however, has the highest concentration of Abenaki Indians and Indian/European admixture of any county in the state. The Abenaki and Abenaki/French Canadian population are a chronically poor and underemployed sector in rural Franklin County and across the border in the St. Jacques River region of Canada, where they have been since the Western Abenaki were resettled in the eighteenth century. The victim, like the accused, was half Abenaki, half French-Canadian and was assaulted where she lived, in a trailer park, about one third of whose residents are of Abenaki ancestry. It is a fair presumption that a large fraction of the victim's circle of acquaintance came from the Indian population. No information exists on the frequency of DNA patterns among Abenaki and Iroquois, and on this basis the judge excluded the DNA evidence. But the state could easily argue that a trailer park is open to access from any passer-by and that the general population of Vermont is the appropriate base of comparison. Rather than objective science we are left with intuitive arguments about the patterns of people's everyday lives.

The dream of the prosecutor, to be able to say, "Ladies and gentlemen of the jury, the chance that someone other than the defendant could be the criminal is 1 in 3,426,327" has very shaky support. When biologists have called attention to the weaknesses of the method in court or in scientific publications they have been the objects of considerable pressure. One author was called twice by an agent of the Justice Department, in what the scientist describes as intimidating attempts to have him withdraw a paper in press.[6] Another was asked questions about his visa by an FBI agent attorney when he testified, a third was asked by a prosecuting attorney how he would like to spend the night in jail, and a fourth received a fax demand from a federal prosecutor requiring him to produce peer reviews of a journal article he had submitted to the *American Journal of Human Genetics*, fifteen minutes before a fax from the editor of the journal informed the author of the existence of the reviews and their contents. Only one of our authors, Christopher Wills, discusses the forensic use of DNA, and he has been a prosecution witness himself. He is dismissive of the problems and seems to share with prosecutors the view that the nature of the evidence is less important than the conviction of the guilty.

Both prosecutors and defense forces have produced expert witnesses of considerable prestige to support or question the use of DNA profiles as a forensic tool. If

---

6. Pressure against the paper was also brought by scientists in the genome sequencing establishment on the editor of the journal in which it was to be published, including one of the contributors to *The Code of Codes*. As a result the editor delayed its publication, demanded changes in galley proof, and asked two defenders of the method to write a counterattack. One report of the scandal is given in Lesley Roberts's "Fight Erupts over DNA Fingerprinting," *Science*, December 20, 1991, pp. 1,721–1,723.

professors from Harvard disagree with professors from Yale (as in this case), what is a judge to do? Under one legal precedent, the so-called "*Frye* rule,"[7] such a disagreement is cause for barring the evidence which "must be sufficiently established to have gained general acceptance in the particular field in which it belongs." But all jurisdictions do not follow *Frye*, and what is "general acceptance," anyway? In response to mounting pressure from the courts and the Department of Justice, the National Research Council (NRC) was asked to form a Committee on DNA Technology in Forensic Science, to produce a definitive report and recommendations. They have now done so, adding greatly to the general confusion.[8]

Two days before the public release of the report, *The New York Times* carried a front-page article by one of its most experienced and sophisticated science reporters, announcing that the NRC Committee had recommended that DNA evidence be barred from the courts. This was greeted by a roar of protest from the committee, whose chairman, Victor McKusick of Johns Hopkins University,

---

7. Based on *Frye* v. *United States* 293 F. 2nd DC Circuit 1013, 104 (1923).

8. DNA *Technology in Forensic Science*, report of the Committee on DNA Technology in Forensic Science (National Academy Press, 1992). The reader should know that I am not a disinterested party either with respect to the report or to the body that sponsored it. I have twice testified in federal court on the weaknesses of DNA profiles, am the author of a position paper that was a basis for the original very critical version of the NRC report's chapter on population considerations, and am the author, with Daniel Hartl, of a highly critical paper in *Science* that was the object of considerable controversy. I resigned from the National Academy of Sciences in 1971 in protest against the secret military research carried out by its operating arm, the National Research Council.

held a press conference the next morning to announce that the report, in fact, approved of the forensic use of DNA substantially as it was now practiced. The *Times*, acknowledging an "error," backed off a bit, but not much, quoting various experts who agreed with the original interpretation. A member of the committee was quoted as saying he had read the report "fifty times" but hadn't really intended to make the criticisms as strong as they actually appeared in the text.

One seems to have hardly any other choice but to read the report for oneself. As might be expected the report says in effect, "none of the above," but in substance it gives prosecutors a pretty tough row to hoe. Nowhere does the report give wholehearted support to DNA evidence as currently used. The closest it comes is to state:

> The current laboratory procedure for detecting DNA variation . . . is *fundamentally* sound [emphasis added].
>
> . . .
>
> It is now clear that DNA typing methods are a most powerful adjunct to forensic science for personal identification and have immense benefit to the public.

and further that

> DNA typing is capable, *in principle*, of an extremely low inherent rate of false results [emphasis added].

Unfortunately for the courts looking for assurances, these statements are immediately preceded by the following:

> The committee recognizes that standardization of practices in forensic laboratories in general is more problematic than in other laboratory settings; stated succinctly, forensic scientists have little or no control over the nature, condition, form, or amount of sample with which they must work.

Not exactly the ringing endorsement suggested by Professor McKusick's press conference. On the other hand there are no statements calling for the outright barring of DNA evidence. There are, however, numerous recommendations which, taken seriously, will lead any moderately businesslike defense attorney to file an immediate appeal of any case lost on DNA evidence. On the issue of laboratory reliability the report says:

> Each forensic-science laboratory engaged in DNA typing must have a formal, detailed quality-assurance and quality-control program to monitor work.

and

> Quality-assurance programs in individual laboratories alone are insufficient to ensure high standards. External mechanisms are needed.

. . .

> Courts should require that laboratories providing DNA typing evidence have proper accreditation for each DNA typing method used.

The committee then discusses mechanisms of quality control and accreditation in greater detail. Since no laboratory currently meets those requirements and no accreditation agency now exists, it is hard to see how the committee's report can be read as an endorsement of the current practice of presenting evidence. On the critical issue of population comparisons the committee actually uses legal language sufficient to bar any of the one-in-a-million claims that prosecutors have relied on to dazzle juries:

> Because it is impossible or impractical to draw a large enough population to test directly calculated frequencies of any particular profile much below 1 in 1,000, there is not a sufficient body of empirical data on which to base a claim that such frequency calculations are reliable or valid.

"Reliable" and "valid" are terms of art here and Judge Jack Weinstein, who was a member of the committee, certainly knew that. This sentence should be copied in large letters and hung framed on the wall of every public defender in the United States. On balance, *The New York Times* had it right the first time. Whether by ineptitude or design the NRC Committee has produced a document rather more resistant to spin than some may have hoped.

In order to understand the committee's report, one must understand the committee and its sponsoring body. The National Academy of Sciences is a self-perpetuating honorary society of prestigious American scientists, founded during the Civil War by Lincoln to give expert advice on technical matters. During the Great War, Woodrow Wilson added the National Research Council as the operating arm of the Academy, which could not produce from its own ranks of eminent ancients enough technical competence to deal with the growing complexities of the government's scientific problems. Any arm of the state can commission an NRC study and the present one was paid for by the FBI, the NIH Human Genome Center, the National Institute of Justice, the National Science Foundation, and two non-federal sources, the Sloan Foundation and the State Justice Institute.

Membership in study committees almost inevitably includes divergent prejudices and conflicts of interest. The Forensic DNA Committee included people who had testified on both sides of the issue in trials and at least two members had clear financial conflicts of interest. One was forced to resign near the end of the committee's deliberations when the full extent of his conflicts was revealed. A preliminary version of the report, much less tolerant of DNA profile methods, was leaked to the FBI by two members of the committee, and the Bureau made strenuous representations to the committee to get them to soften the offending sections. Because science is supposed to find objective truths that are clear to those with expertise, NRC findings do not usually contain majority and minority reports, and, of course, in the present case a lack of unanimity would be the equivalent of a negative verdict. So we may expect reports to contain contradictory compromises among contending interests, and public pronouncements about a report may be in contradiction to its effective content. *DNA Technology in Forensic Science* in its formation and content is a gold mine for the serious student of political science and scientific politics.

There is no aspect of our lives, it seems, that is not within the territory claimed by the power of DNA. In 1924, William Bailey published in *The Washington Post* an article about "Radithor," a radioactive water of his own preparation, under the headline, "Science to Cure All the Living Dead. What a Famous Savant has to Say about the New Plan to Close Up the Insane Asylums, Wipe Out Illiteracy, and Make over the Morons by his Method of Gland Control."[9] Nothing was more up-to-date in the 1920s than a combination of radioactivity and glands. Famous savants, it seems, still have access to the press in their efforts to sell us, at a considerable profit, the latest concoction.

---

9. See M. Allison, "The Radioactive Elixir," *Harvard Magazine*, January–February 1992, pp. 73–75.

Epilogue

The promise of great advances in medicine, not to speak of our knowledge of what it is to be human, is yet to be realized from sequencing the human genome. Although the DNA carrying the normal form of a gene has been put into the bodies of people suffering from a variety of genetic disorders, there is not a single case of successful gene therapy in which a normal form of a gene has become stably incorporated into the DNA of a patient and has taken over the function that was defective. There was, for example, an early report that normal DNA sprayed into the lungs of a cystic fibrosis patient was taken up by cells and resulted in partial recovery, but the optimism was premature. An alternative method has been to graft genetically normal cells or tissue into a patient in the hope that the cells will proliferate and take over normal function. A case has been reported of a considerable lowering of cholesterol level in a patient suffering from an extreme form of inherited hypercholesterolemia after liver cells with the normal form of the gene were implanted. Unfortunately the lowered level was still pathologically high and we await news of further progress. There seems no fundamental reason why such methods should not work sometimes, but the trick has not yet been discovered. Over and over, reports of first isolated successes of some form of DNA therapy appear in popular media, but the prudent reader should await the *second* report before beginning to invest either psychic or material capital in the proposed treatment.

One of the issues raised around the original Human Genome Project was that it seemed to pay no attention to the known genetic variation from individual to individual and from group to group. Whose genome was going to be represented in the human genome? As a result of agitation around this issue a small fraction of the budget of the project was diverted to studying genetic variation. One outcome was the formation of the Human Genome Diversity Project, a cooperative project of a number of human geneticists led by L. L. Cavalli-Sforza of Stanford University, to characterize genetic variation across the species. Originally the intent was to obtain a picture of the genetic patterns in a great diversity of small or disappearing populations, but it was protested that such a study was all very well for anthropologists but not for a random sample of humanity who are mostly living in the densely populated regions. As a result the project now plans to sample more indiscriminately.

But even then the main problems posed for the genome project by genetic polymorphism are not solved. We will still not know whether the bit of genome sequenced from a particular donor carries one copy of a defective sequence. We will still not know, from comparing sequences from a large number of sick and well people, which of the many nucleotide differences between them is responsible for the abnormality. That is not to say that the Diversity Project is useless. It will greatly increase the observed repertoire of DNA sequences carried by well and by sick people and so help us to avoid being led astray from too narrow a base of comparison. For example, there are over two hundred different nucleotide changes, any of which can cause hemophilia. Most of these have been discovered by sequencing the relevant gene in people from different regions of the world. The genetic array of hemophilias in Calcutta is not the same as in Germany. Thus, the study of diversity will provide us with the raw material we need to understand what makes a hemophiliac, but in the end the molecular biology of the gene and protein must be explored. That is, we need to understand how the different nucleotide changes cause a deficiency or absence of the needed clotting protein, or, if the protein is present but abnormal in its structure, how such structural variation interferes with the clotting reaction. Knowing that a gene variation is at the root of disorder is useless unless it is possible to provide a story of physical mediation that can be translated into therapeutic action.

The main developments in genome research have revolved around the generation of the sequence information itself, and the application of that information to the production of pharmaceutical treatments. Just as for cloning, the course of human genome research in the last half-dozen years cannot be understood outside the context of commercial interest.[10]

The Human Genome Project, funded by the NIH and the Department of Energy, has so far sequenced about 4 percent of the entire three billion bases in human DNA, but the rate is accelerating so that the latest of the

---

10. See the epilogue to Chapter 8.

constantly revised estimates puts the completion of the project in 2003. However, there is now commercial competition. Early in the project Craig Venter, one of the cleverer participants, fell out with the directors over a strategic issue. Of the three billion nucleotides in the human genome, it is estimated that only about 5 percent are really in genes that code for proteins used by the organism. The remaining 95 percent are said to be in "junk" DNA without function, which is to say that nobody happens to know if it has a function. If it really is junk then, as Venter pointed out not unreasonably, sequencing it should be a secondary objective for a project whose claim to legitimacy is to cure human disease and understand human nature. He proposed that the Human Genome project could save a lot of time and money using a method of his invention that would pick out only the genic DNA. When the directors of the project disagreed, he quit and set up in business for himself.

Venter has now changed his mind about what is worth doing. His Institute for Genomic Research recently joined with a scientific-instrument producer, the Perkin-Elmer Corporation, to sequence the entire genome, junk and all, using hundreds of newly designed automated sequencers. As yet, none of these machines has actually come off the assembly line, but when they do they will become available on the open market for a mere $300,000 apiece. The total projected cost is only about $250 million and the total time needed was originally estimated to be three years, if the robots really work. In March 1999 the competition between the public and private sequencing projects was intensified by the announcement that the public project intended to finish 90 percent of the sequence by the spring of 2000, while Venter's timetable is still aiming at completion in the middle of 2001.

There is a good deal more at stake than the profit from some machines or a contract to determine the sequence. Since the early 1990s the courts have held that a gene sequence is patentable, even though it is a bit of a natural organism. (At the end of 1998, the CEO of one genome company, Human Genome Sciences, a former professor at the Harvard Medical School, wrote that his corporation had filed over 500 gene patent applications.[11]) One value of a patent on a gene sequence lies in its importance in the production of targeted drugs, either to make up for the deficient production from a defective gene or to counteract the erroneous or excessive production of an unwanted protein. In the first instance, the protein coded by the gene may itself be the drug, in which case it can be produced by transferring the gene to a bacterium or other cell, and growing the protein in mass quantities in fermenters. A classic example is the production of human insulin to supplement the lack of its normal production in diabetics. Alternatively, the cell's production of a protein coded by a particular gene, or the physiological effect of the genetically encoded protein, could be affected by some molecule synthesized in an industrial process and sold as a drug. The original design of this drug and its

---

11. William A. Haseltine, "Life by Design: Gene Mapping, Without Tax Money," *The New York Times*, May 21, 1998, p. A33.

ultimate patent protection will depend upon having rights to the DNA sequence that specified the protein on which the drug acts. Were the patent rights to the sequence in the hands of a public agency like the NIH, a drug designer and manufacturer would have to be licensed by that agency to use the sequence in its drug research, and even if no payment were required the commercial user would not have a monopoly, but would face possible competition from other producers.

A promising case of a drug developed from a knowledge of the genetic control of protein synthesis is Herceptin, registered, produced, and marketed by Genentech for the treatment of breast and ovarian cancers. One form of these cancers is the consequence of the duplication of the HER-2 gene, which results in the overproduction of a protein that greatly stimulates cell division. Herceptin is an antibody molecule that specifically blocks this stimulation of cell division. It remains to be seen how profitable Herceptin will be, but the present value of possessing such a drug is estimated at about five billion dollars.[12]

There are currently ten genomics companies involved in possible drug production, in collaboration with major pharmaceutical companies. None has yet made any money by selling a drug based on genome sequencing, but their prospectuses all predict a profit soon. Before a pharmaceutical company can make any money on the

---

12. This is calculated as the present value of the drug when the after-tax income stream is projected for thirty-five years using a long-term estimate of average rates of return (Genomics II, Lehman Brothers, January 23, 1998).

production and sale of a drug, clinical trials must satisfy both medical practitioners and the FDA that a drug is both effective and safe, and even then the costs of production and marketing may exceed what can be taken in. There is also the possibility of commercial success in diagnostic testing, but it remains in the future. For example, using the DNA sequence, a test has been developed for the BRCA1 mutation that is involved in a small fraction of breast cancers. Despite a great deal of publicity for the test, its owner, Myriad Genetics, has yet to show a profit.

It may turn out, in the end, that the providers of capital have been as deluded by the hype of the human genome as has anyone else. Judging by the results so far the prudent investor may be better off spending a week at Saratoga. Only a foolhardy person would predict that no gene therapies will ever be commercially successful. Even at Saratoga long shots pay off once in a while.

It was impossible to say in 1992 how far the Human Genome Project or drug therapies based on it would have developed in seven years. What became clear very quickly, however, was the future of the forensic applications of DNA technology. The report of the National Academy of Sciences was doomed to the dustbin. At first the Department of Justice and other law enforcement agencies were quite happy with the report because it gave a generalized approval, in principle, to the use of DNA profiles in identification. But more and more courts began to rule DNA evidence inadmissible when the detailed analysis of the report was brought out at trials. The problem of genetic differences between ethnic groups that I describe in my review was particularly damaging to prosecution calculations of how likely the crime scene DNA was to match a random innocent person. It soon became obvious that prosecutorial agencies were going to press for some action that would validate DNA evidence. And so they did. The National Academy of Sciences, through its subsidiary, the National Research Council, is obliged to carry out any inquiry for which it has competence when that inquiry is requested and paid for by an entity of the federal government. The result is that it is sometimes asked to visit the same question again if the clients are not satisfied with the first outcome. The most notorious case was a report showing that high-protein dog food was bad for pets' kidneys, a result that was unsatisfactory to a leading producer then engaged in a high-pressure advertising campaign for its high-protein dog diet. The political clout of the dog food cannery was sufficient that three successive reports were called for, all unsatisfactory, before the company and their government representative finally gave up.

With the dog food case as a procedural precedent, the director of the FBI asked for a new report on forensic DNA in 1993 and money was also provided by other interested agencies. There was no great problem in predicting the outcome of the committee's deliberations once the membership was known, since by 1993 everyone in the field had expressed a clear view on the matter. I wrote to the President of the Academy offering to save everyone a lot of time and money by writing the report if he would just send me the list of the committee members, but he did not take my suggestion. Before the committee had even met, the chairman, an eminent geneticist, gave a speech at a meeting of the Forensic Science Association, in which he assured the representative of the FBI that everything would turn out well. The two main issues of contention, quality control of crime laboratories and the difficulties that laypersons have in understanding probability statements, were neatly finessed in the report. All laboratories that sequence DNA have problems of cross-contamination between samples. This becomes particularly acute when a minute sample of DNA, say from a scraping of a bit of dried blood, is to be compared against a large sample of blood taken from a suspect. If these are not handled with great care and attention, DNA from the large sample may wind up contaminating the small one. Moreover, a lot of DNA comparison is not done in the relatively sophisticated central crime laboratory of the FBI, but in local state and county forensic facilities. The FBI laboratory itself had repeatedly refused to allow independent assessors to observe their procedures or submit to blind proficiency tests. Yet the best the committee could recommend was that "laboratories should adhere to high quality standards ... and make every effort to be accredited for DNA work."[13] Well, perhaps not *every* effort.

As to the problem of juror comprehension of

---

13. National Research Council, *The Evaluation of Forensic DNA Evidence* (National Academy Press, 1996).

probability statements, the recommendation was that "behavioral research should be carried out to identify any conditions that might cause a trier of fact to misinterpret evidence on DNA profiling and to assess how well various ways of presenting expert testimony on DNA can reduce any such misunderstandings." This recommendation neatly ignores the already extensive literature showing that laypersons often misunderstand probability statements even when they are presented in a one-to-one interview. So, for example, studies funded by the NIH of the results of genetic counseling found that couples who were told that they had one chance in four of having an affected child would sometimes respond that they were not worried because they were only having two children.

As might be expected, with the new report in hand prosecutorial agencies no longer have problems in court with the admissibility of DNA evidence.

Second Epilogue

After the Genome, What Then?

On Monday, February 12, 2001, *The New York Times*, on its front page above the fold, leaked the news that the two competing projects to sequence the human genome were about to announce on that very day that they had indeed located the Holy Grail. Then, on Thursday and Friday, the scientific papers giving the details appeared, surrounded by a penumbra of commentary, analysis, and promises of a rosy future for human health and self-knowledge. It might seem remarkable that both publicly funded and commercial projects should have independently accomplished their ends of sequencing the three billion nucleotides of the human genome, analyzing the sequence, and publishing their findings within a day of each other, but it was no coincidence. It was, in fact, the carefully prearranged and orchestrated outcome of a truce between the contenders announced at a joint press conference the previous June.

Their decision that the human DNA sequence was now definitively determined was an arbitrary one since there are admitted to be gaps amounting to about 6 percent of the sequence yet to be filled in. As in long-standing political struggle, the exhausted parties simply decide that enough is enough, but, as in political cases, some occasional sniper fire is still heard. So, Celera Genomics' commercial project claims that its sequence is more accurate than the publicly funded one, while the International Human Genome Sequence Consortium claims that only by use of their publicly available intermediate results could Celera have assembled their sequence in the first place.

A small irony of the simultaneous publication is that the public project, supported in large part by American government funds, used as its vehicle the English commercial scientific journal *Nature*, owned by Macmillan, while the commercial Celera project used *Science*, the organ of the American Association for the Advancement of Science, a nonprofit professional society. Some of the details of publication are immensely revealing of the sociology of science and scientific writing. Modern natural scientific work often requires the joint efforts of several professional participants, all of whom depend on publication for their career advancement and the acquisition of further research funding. The result has been the dominance of the jointly authored scientific report. The most recent issue of *Genetics*, the major international publication in the field, contains forty-one research papers, none of which was the product of only a single author.

As befits the monster human genome sequencing projects, their author lists are monsters: 275 authors for the commercial project and 250 for the International Consortium (both lists being characterized as "partial"). The order of authors is generally also of great import in the acquisition of scientific credit and the decoding of these lists would be an interesting exercise for sociologists. Aside from the predictable first authorship of Craig Venter, the head of Celera, on the commercial publication and of Eric Lander, the director of the Whitehead Institute, which was responsible for more sequencing than the other cooperators in the public project, it is not obvious how we are to understand the order of authors. Nor is there any hint of who, among hundreds of "authors," actually wrote the papers. This too is a revelation of the assumptions of scientific work. Scientists, by their practices, seem to place little importance on the actual composition of their communications. For example, they never read written papers aloud when they give talks about their work, but speak ex tempore. For other intellectuals the words are the matter, but scientists think of themselves as simply reporting objectively the facts of nature. Like the Delphic Oracle, they sit perched on their tripods, with upturned eyeballs, and out of their mouths issue nature's words. But, of course, the long reports on the human genome, like any scientific report, are filled with analysis and interpretation all informed by communal and individual judgments of what is significant and what is to be ignored.

And what is significant in the human genome sequence? The major irony of the sequencing of the human genome is that the result turns out not to provide the answer to the chief question that motivated the project. Now that we have the complete sequence of the human genome we do not, alas, know anything more than we did before about what it is to be human. At the time of the completion of the human genome sequence, scientists already knew the complete DNA sequences of thirty-nine species of bacteria, a yeast, a nematode worm, the fruit fly, *Drosophila*, and the mustard weed, *Arabidopsis*. In each case it is possible to estimate how many genes are present in the genome, using two methods. The first is to compare stretches of DNA sequence with sequences of particular genes already known from a variety of organisms. The other, for DNA that does not match already known genes, is to use certain sequence motifs that are common to all genes. When this so-called "annotation" of the human genome was done it was estimated that humans have about 32,000 genes. This seems a rather small number when the comparison is made with the fruit fly (13,000), the nematode worm (18,000), and the mustard weed (26,000). Can human beings really only have 75 percent more genes than a tiny worm and a mere 25 percent more than a weed? If, as the eminent molecular biologist Walter Gilbert wrote, a knowledge of the human genome would cause "a change in our philosophical understanding of ourselves," that change has not been quite what was hoped for. It appears that we are not much different from vegetables, if we can judge from our genomes.

The reaction to the discovery that human beings do not have much more genomic information than plants and worms has been to call for a new and even more grandiose project. It is now agreed among molecular biologists that the genome was not really the right target and that we now need to study the "proteome," the complete set of all the proteins manufactured by an organism. Surely the very complex human being must have many more different proteins than a small flowering plant. Although the devotees of the genome project kept assuring us that genes made proteins and therefore when we had all the genes we would know all the proteins, they now say that, of course, they knew all along that genes don't make proteins. Genes only specify the sequence of amino acids that are linked together in the manufacture of a molecule called a polypeptide, which must then fold up to make a protein. But there are many different ways in which a long polypeptide can fold, resulting in different proteins. The way in which the folding occurs may be different in different cells of different organisms and depends in part on the presence of small molecules, like sugars, and on other proteins.

Moreover, a gene is divided up into several stretches of DNA, each of which specifies only part of the complete sequence in a polypeptide. Each of these partial sequences can then combine with parts specified by other genes, so that from only a few genes, each made up of a few subsections, a very large number of combinations of different amino acid sequences could be made by mixing and matching. So knowing all the genes of a human being doesn't really tell us what we want to know.

One prominent opponent of the genome sequencing project, William Haseltine, CEO of Human Genome Sciences, has long claimed that the right way to find all the human genes is not to sequence the genome itself, but to go directly to the products that the cell makes when it reads the genome. These products, nucleotide sequences called "messenger RNAs," are then used by the cell to manufacture the polypeptides. Haseltine claims to have detected 90,000 of these messengers in human cells, but whether that means there are 90,000 different genes or 90,000 different combinations of bits and pieces from approximately 32,000 genes is unclear, given that no detailed accounts of his findings have been published.[1]

The call for a proteome project comes just in time to solve the practical problem created by the completion of the genome project. What is Big Science going to do now? A proteome project will be much larger than the genome project and will take much longer to finish. There are, we suppose, a lot more different proteins than there are genes. Moreover, the sequencing of the DNA of a gene is technologically trivial in comparison with the determination of the three-dimensional structure of a protein. In the past it would take a Ph.D. candidate three years to determine the sequence of a single protein. New automated technologies are being developed, but the proteome project will be guaranteed to occupy a large number of scientist-years well into the future.

As interest shifts from genes to proteins, so the promises of cures for all of our ills will shift from genome fixes to protein fixes. The special Human Genome issues of *Science* and *Nature* already prefigure

---

1. See the story about Human Genome Sciences in *The Financial Times* (London), June 12, 2001, p. 15.

this change. Amid the many articles of the standard sort like "Toward Behavioral Genomics" and "Cancer and Genomics" is one called "Proteomics in Genomeland," and one, "Dissecting Human Disease in the Post-Genomic Era," which describes the shift from genomics to proteomics as one of the "Paradigm Shifts in Biomedical Research." As yet the promise that the study of DNA sequences will lead to cures for illness has remained unfulfilled for any human disease, although some gene-based drugs are undergoing clinical trials. Proteomics has arrived in the nick of time to assume the burden, and with more reason. The historical successes of molecular medicine have been precisely in developing drug therapies, dietary regimes, or substitute sources of faulty or missing proteins. The provision of insulin to diabetics and the amelioration of at least the most debilitating symptoms of the inherited metabolic disease Phenylketonuria (PKU) by dietary restriction are the best-known examples.

The subject of DNA seems filled with ironies. The struggle over the forensic use of DNA profiles to link defendants to crime scenes is now over. The use of such evidence is now routine despite the fact that the problems posed by the presentation of quantitative probability arguments to juries and the lack of uniform rigorous quality control of laboratory work have never been resolved; nor is there any effort being made to deal with these issues. The cessation of that struggle is partly a result of the feeling on the part of those who originally opposed the introduction of DNA evidence that the battle was unwinnable. The second National Academy of Sciences report placed the full weight of the scientific establishment behind the use of DNA profiles and a series of court decisions have validated their admissibility as evidence.[2]

But the frustrated opponents of DNA profiles introduced as incriminating evidence have partly made use of the legitimation of the technique to turn it to the opposite purpose. People who, on the basis of eye-witness identification or circumstantial evidence, were convicted of violent crimes like rape and murder and who were given long sentences or threatened with execution are now being released on the basis of their DNA profiles. In many cases physical evidence in the form of blood or semen samples recovered from the crime scene or victim have been preserved. When these are subsequently compared with the DNA profile of the convicted person and a mismatch is found, then that person is definitively exonerated.

The word of these successful rescues from prison and execution has spread and the demand on the part of prisoners for reopening of their cases has grown enormously. Prosecutorial forces have resisted these demands as strongly as they can, fearful of a deluge of reconsiderations of their successful past prosecutions, and only a few defense attorneys have the resources to take up these old cases again in the face of the strong

2. The first report of the National Academy was DNA Technology in Forensic Science (National Academy Press, 1992). The second was the National Research Council's The Evaluation of Forensic DNA Evidence (National Academy Press, 1996).

resistance on the part of the state. But there have been some notable successes.

The leaders of the movement to use DNA evidence for exculpations have been the attorneys Peter Neufeld and Barry Scheck, experts on forensic uses of DNA. Using the resources and fame they acquired in their successful defense of O. J. Simpson, they have organized the Innocence Project, which, together with other efforts inspired by it, has thus far succeeded in obtaining the release of more than ninety prisoners serving long-term sentences or living under the threat of execution. Unfortunately, the necessary physical evidence has often not been preserved, and when it has, considerable effort of time and money is needed to obtain access to it, so the Innocence Project is not likely to lead to a wholesale reconsideration of past convictions. Prosecutors, however, can be counted on to make increasing use of DNA evidence to secure convictions in order to protect those convictions against challenges. Moreover, the demonstration that innocent people have been sentenced to death has given opponents of the death penalty a very powerful argument.

—*The New York Review of Books*, July 19, 2001

Chapter 6

WOMEN VERSUS THE BIOLOGISTS

"Women Versus the Biologists" was first published in The New York Review of Books *of April 7, 1994, as a review of* Exploding the Gene Myth: How Genetic Information Is Produced and Manipulated by Scientists, Physicians, Employers, Insurance Companies, Educators, and Law Enforcers, *by Ruth Hubbard and Elijah Wald (Beacon, 1993);* Biological Woman—The Convenient Myth, *edited by Ruth Hubbard, Mary Sue Henifin, and Barbara Fried (Schenkman Publishers, 1982);* Women's Nature: Rationalizations of Inequality, *edited by Marian Lowe and Ruth Hubbard (Pergamon Press/Teachers College Press, 1983);* Pitfalls in Research on Sexual Gender, *edited by Ruth Hubbard and Marian Lowe (Gordian Press, 1979);* The Politics of Women's Biology, *by Ruth Hubbard (Rutgers University Press, 1990); and* The Shape of Red: Insider/Outsider Reflections, *by Ruth Hubbard and Margaret Randall (Cleis Press, 1988).*

1.

THE CENTRAL SOCIAL agony of American political and social life since the founding of the Republic has been caused by the problem of equality. Our domestic political history has been dominated by the demand for equality and the resistance to that demand. A destructive civil war, urban riots, the burning of cities, major legislation and judicial struggles, and the local social and political structures of a large section of the United States have all, at least at the level of public consciousness, been responses to the manifest inequality of status, wealth, and power in a society whose chief claim to legitimacy has been its devotion to equality.

Western history had, of course, always been marked by civil wars, peasant uprisings, rick burnings, machine breakings, and urban riots, but these were in the name of bread, land, and work. The demand for social and political equality was a creation of the ideologues of modern society. Both in Europe and North America, the bourgeois revolutions of the seventeenth and eighteenth centuries, which overthrew the *ancien régime* of restricted privilege, were based on the slogans of *liberté, égalité, fraternité.* "All men are created equal, and they are endowed by their Creator with certain unalienable rights."

These are the slogans of our childhood, the unquestioned assertion of the basis of our political and civil life. Yet the facts of that life are in direct contradiction with the ideology. It is obvious to everyone, no matter how optimistic their politics, that there are immense inequalities of social status, power, and wealth among individuals, among races, between the sexes. While Jefferson could not have meant what he said about *all* men being created equal, since a mere ten years later the framers of the Constitution arranged that slaves would be counted as only three-fifths a person, he meant literally all *men* since the political rights of women were not established for another 130 years.

The social tension created by the contradiction between the ideals of equality and the manifest existing inequalities has been, in part, relieved by institutional and judicial arrangements. Constitutional amendments, Supreme Court decisions, civil rights legislation have all been devoted to creating a better fit between the ideal and the real. Yet major inequalities remain, and it does not seem that further judicial and institutional changes of a radical kind will be accomplished. The movement for women's rights, in particular, is stalled and even reversed, with the Equal Rights Amendment no longer a political issue and abortion rights in at least partial retreat. What is the solution? The Enlightenment, having created the problem in the first place by the claim for individual rights, also provided a tool for legitimizing inequality through its implied claims that the individual is supremely responsible for causing the unequal situation he or she occupies.

Accompanying the static and unchanging social position in which prerevolutionary Europeans found themselves was the view that divine causation provided legitimacy to hierarchal society. The doctrine of Grace was the guarantor of social stability, and only on those occasions when Divine Grace was conferred or withdrawn could one expect to change one's social position. (Cromwell observed that although Charles I ruled *Dei Gratia*, Grace had been removed from him as evidenced by his severed head.) In the postrevolutionary world, individuals are said to acquire their position in society by their own efforts, and these efforts must be effective if the society built on them is to be legitimate. Individuals are ontologically prior to the collectivity in this worldview, and so the properties of society are simply the accumulated consequences of the properties of individuals. Whether it be Hobbes's derivation of the war of all against all from the self-expansive properties of individuals in a world of limited resources, or Weber's view of the supreme importance to human institutions of outstanding leaders like Bismarck, or Durkheim's notion of the collective mind of society, the properties of the individual human being become, for modern social theory, the determinant of social relations. "Natural rights for natural men" has replaced "*Dieu et mon Droit.*"

If, despite our best institutional efforts to destroy artificial barriers to entry, blacks as a group continue to have lower social and economic status than whites, then we must look into the properties of blacks as individuals for the causes of that inequality. If women lack power, it must be that women are the weaker sex. But if the properties of society are the properties of individuals writ large, then the study of society must become the study of individuals, for social causes are, ultimately, individual causes. To understand the origin and maintenance of social structures, we must, in this view, understand the ontogeny of individuals. Thus political economy becomes applied biology. Economics becomes the study of consumer psychology, worker incentives, and investor behavior, of individual utilities in two-person, zero-sum games.

The first serious "scientific" study of the internal biological causes of social position was Cesare Lombroso's late-nineteenth-century criminal anthropology, which claimed that criminals were born and not made. This theory of innate criminality, updated to implicate faulty DNA, has a modern current and, indeed, is taught at Harvard. There has been, since Lombroso, a major intellectual industry tracing the causes of social inequality between classes, races, and the sexes. A vast literature has been created and, in reaction, a smaller group of debunking critics of biological determinism has emerged[1] in what those of us involved liken to the work of a volunteer fire department. No sooner has

---

1. See, for example, S. J. Gould, *The Mismeasure of Man* (Norton, 1981), reviewed in Chapter 1, and R. C. Lewontin, S. P. R. Rose, and L. J. Kamin, *Not in Our Genes* (Pantheon, 1984).

one blaze set by intellectual incendiaries been doused
by the cool stream of critical reason than another
springs up down the street.

There is, at present, no aspect of social or individual
life that is not claimed for the genes. Richard Dawkins's[2] claim that the genes "create us, body and mind"
seemed the hyperbolic excess of a vulgar understanding
in 1976, but it is now the unexamined consensus of
intellectual consciousness propagated by journalists
and scientists alike. The belief in the absolute primacy
of the internal over the external is nowhere more manifest than in the demands of the biological parents of
Jessica DeBoer and Kimberly Mays to assert their
genetic rights over the lives of their children who had
been raised by others. Every physical, psychic, or social
ill, every perturbation of the body corporeal or politic
is said to be genetic. There are, according to "scientific
studies," genes for schizophrenia, genes for sensitivity
to industrial pollutants and dangerous workplace conditions, genes for criminality, genes for violence, genes
for divorce, and genes for homelessness. While there
have been a few essays and reviews questioning this
genomania or at least considering its claims with some
measure of skepticism,[3] there has been no generally
critical book on the diverse claims for the power of DNA
until the appearance of *Exploding the Gene Myth* by
Ruth Hubbard and her son, Elijah Wald, who provided

2. Richard Dawkins, *The Selfish Gene* (Oxford University Press, 1976).

3. See Chapter 5.

the rhetorical skills to supplement Hubbard's biological expertise.

Past books on eugenics or on biological determinism in general have discussed the claims historically and ideologically, attempting to explain the rise in biologistic explanations as political and ideological phenomena, but expending little effort on exposing the biological issues themselves. *Exploding the Gene Myth* puts to one side these political and ideological forces and concentrates on describing the "gene myth," providing an accessible account of what is really known about the relevant biology of reproduction, and discussing the social and legal consequences of the reliance on genetic explanations and causes.

Exploding the Gene Myth begins with a brief survey of how claims about genetics and actual medical and social practice based on genetics affect our lives. The structure of the authors' description follows the lines laid down by Daniel Kevles in his extremely influential book on the history of human genetics.[4] Kevles argued convincingly that the eugenics movement, having been discredited as a movement for social improvement, largely by the extreme racism of the Nazis, was converted into human clinical genetics, whose object is not to better society as a collective but to provide to individuals and families diagnosis, counseling, and therapies to alleviate individual suffering.

---

4. Daniel J. Kevles, *In the Name of Eugenics: Genetics and the Uses of Human Heredity* (Knopf, 1985).

By extension, diagnosis and counseling, but certainly not therapies, are provided to employers and insurers to screen out workers who are potential health risks. The explanatory model of human disorders provided by genetics is based on the claim that genes determine significant aspects of human anatomy, physiology, and behavior. Genes are said to "control," "create," or "determine" the physical and psychic development of individuals, because the DNA is a set of instructions to the biochemical processes of the cells that make us up.

"Normal" individuals, then, have normal genes, while a very large fraction of the sick (including those with heart disease and cancer) owe their diseases to abnormal sequences of DNA. The first problem of human genetics, then, is to identify the gene "for" an abnormality, and provide a procedure for recognizing its presence in an individual. Carriers of defective heredity can then be advised on a course of preventive maintenance, or a therapy that may in the future include the actual replacement of the defective gene by a normal component, rather like the replacement of a bad steering mechanism in a manufacturer's recall of a car. At the worst, having no therapy to offer, the geneticist can warn the carrier of defective DNA that it is time to make her will.

Hubbard and Wald attack this model at its base, by challenging the claim that genes "determine" organisms. They describe, correctly, how the development of an organism is a unique consequence of the interaction of genetic and environmental forces, and always subject to accidents of development. Nor are these accidents that we normally think of as traumata causing birth defects. They are characteristic of every individual life history, for time and chance happeneth to all. Moreover, they explain how, even in cases where genes may play a major role in the causal pathway of a disorder, the model of one gene–one disorder is far too simple. It is by no means clear that diabetes, for example, can be explained by reference to defective genes, but if it can, there must be several or even many genes implicated.

The genetic model of disease leads ineluctably to the disease model of all ills and social deviance. So genetic defects are claimed to lie at the basis of heart disease, schizophrenia, alcoholism, drug dependence, violent behavior, unconventional sex, and shoplifting. Recognizing that any commonsense consideration of these conditions implicates environmental influence, geneticists often refer to "inherited *tendencies*" to these conditions. Hubbard and Wald devote considerable attention to showing how little actual genetic knowledge exists for such "genetic tendencies," and how difficult it is to obtain such knowledge, since the chief tool of investigation is the observed similarity between relatives. The central problem of human genetics is precisely that relatives resemble each other because of both genetic and cultural ties, and we do not know how to disentangle the two.

The sobriety, care, and accuracy of the argument are both its strength and its weakness. No one can accuse

its authors of polemical excesses, ideologically motivated claims, or antiscientific bias. The book can and
ought to be used as a text in law schools and schools of
public policy. The problem is that after the myths are
exploded there is nothing left but a hole in the ground.
The truth about alcoholism, violence, and divorce is
that we don't know the truth. There are no positive
claims about their causes that can be made with any
honest conviction. But saying that our lives are the
consequences of a complex and variable interaction
between internal and external causes does not concentrate the mind nearly so well as a simplistic claim; nor
does it promise anything in the way of relief for individual and social miseries. It takes a certain moral
courage to accept the message of scientific ignorance
and all that it implies.

2.

One place that seems constantly ablaze with fires set
not only by hostile forces from across the tracks but
by the homeowners themselves is the neighborhood
of gender differences. In the struggle for institutional
and legal equality, women have been rather less successful than blacks. The Nineteenth Amendment came
fifty years after the Fifteenth, and in proportion to
their numbers, women are represented among the
CEOs and presidents of large industrial corporations
and major universities in smaller numbers than those

previously excluded minorities, blacks and Jews. When speaking to academic audiences about the biological determination of social status, I have repeatedly tried the experiment of asking the crowd how many believe that blacks are genetically mentally inferior to whites. No one ever raises a hand. When I then ask how many believe that men are biologically superior to women in analytic and mathematical ability, there will always be a few volunteers whose raised hands are accompanied by a snicker or two from the audience and some frowns of disapproval. To admit publicly to outright biological racism is a strict taboo, but the avowal of biological sexism is tolerated as a minor foolishness, unlikely to bring serious consequences.

While white intellectuals have been among the prime opponents of the claims of the biological inferiority of blacks, the struggle against claims of the innate biological inferiority of women has been mostly the work of other women. Partly, this asymmetry is a consequence of the fact that intellectuals, as members of the middle class, have seen racism as having primarily the economic and political consequences of keeping blacks in a permanent underclass, while seeing the disabilities of women as chiefly secondary issues of consciousness, self-esteem, and professional advancement. "Well, it's too bad if women can't be full professors of mathematics or president, but they'll survive."

But it is also a consequence of the ideology of the part of the feminist movement that affirms an essential psychic and cognitive difference between men and women,[5] and that often denies to men the possibility of serving in anything but a supportive role in the fight against claims of the biological inferiority of women. Sometimes these differences are said to be the outcome of the maturing child's relation to its mother, and so are biologically based only at second hand, since it is female biology that prescribes their role as mothers. At other times, it is claimed that psychosocial differences, favorable to women, are directly the consequence of the action of hormones and genes on psychic development. So women have been said to be naturally more cooperative, more loving, less violent and competitive, and more able to conceive problems in broad outline. Of course, it all depends on what one thinks is a desirable trait. The feminist anthropologist Sarah Blaffer Hrdy thinks women are naturally *more* crafty and acquisitive than men, and have been made so by evolution. Instead of criticizing claims of innate and ineluctable psychic differences, such feminists seek to use those claims to the *advantage* of women, and since women have a unique understanding, only they can address the issue on the side of women. That is a mistake that Ruth Hubbard never makes.

No one has been a more tireless and influential critic of the biological theory of women's inequality than

5. See, for example, Elaine Morgan, *The Descent of Women* (Bantam, 1973), for an "esterocentric" view of evolution, and Nancy Chodorow, *The Reproduction of Mothering* (University of California Press, 1978), which stands Freud on his head.

Ruth Hubbard. When the fire brigade is called out to stop the latest arson incident, she can be glimpsed in the smoke directing the hoses, and when the flames are out and everyone has gone home, it is she who undertakes a thorough overhaul of the ashes to keep the glowing embers from rekindling. Hubbard began as a research scientist studying the physiology and biochemistry of vision and became the first and, for a long time, the only woman given a tenured professorship of biology at Harvard in a faculty of forty-three tenured members. Having finally been appointed professor after many years as a research associate, she had the gall to inform her colleagues that she was giving up the research career that had led to her appointment, and would, in the future, devote herself to women's studies and social issues in biology.

The very considerable courage and political conviction required to do this should not be underestimated. As scientists grow older, they often give up research in favor of philosophy, history, or politics, which most younger scientists see as wooly-minded pursuits that do not really require any intellectual rigor. But they do so imperceptibly, pretending always to be involved in scientific work, for only continuing scientific production confers on us the status and ego rewards that we have coveted all our lives. Scientific work creates that bank account of legitimacy which we can then spend on our political and humanist pursuits. To devalue deliberately, in the service of political principle, the past currency of one's life at the very moment when the check has been cashed is not a casual act. Of course, one may criticize the decision on strategic grounds. The extent to which scientists have credibility when they speak about social and political issues depends upon their continual legitimacy as "objective" scientists. By giving up that legitimacy, indeed, by showing a certain disdain for her colleagues' expectations, Hubbard gave up her institutional claim of authority, and thus a certain credibility. "I knew it all the time," her colleagues must have said. "Just like a woman to give up scholarship for nonsense once she gets what she wants." Despite the claim that in the marketplace of ideas it is the better-made product that wins the consumer's heart, it is, in fact, brand loyalty that counts. "Made in Cambridge" has always been worth far more than the force of logic. I once had the occasion to testify as an expert witness, reporting the results of the work of a member of a research team who was a professor of economics. Opposing counsel, when cross-examining me, asked whether Dr. Baker was a professor at *Harvard* (his emphasis). "No," I replied, "at the University of South Carolina." "Aha," he said with a smile and sat down. The defense rested.

3.

The six books reviewed here all bear in one way or another on the problem of the determination of gender differences. Three are collections of essays, including some of her own, that Hubbard has edited together with a chemist, Marian Lowe, and two lawyers engaged in

ecological and women's issues, Mary Sue Henifin and
Barbara Fried, both former students of Hubbard. Two,
The Politics of Women's Biology and *Exploding the Gene
Myth*, are systematic treatments of the relation of inner
biological causes to social identity. The last, *The Shape
of Red*, is an attempt to provide an alternative view of
the development of a woman through an autobiographical exchange with a political comrade, Margaret Randall, who had exiled herself from the United States to
Mexico in response to her sense of alienation from the
direction of American politics, and whose attempt to
return was unsuccessfully resisted by the American government. All are concerned with the relation between
biological subject and biological object, between the
inner and the outer, the individual and the social. And
all, in one way or another, are glosses on that ambiguous slogan of the 1960s, "The personal is political."

The problem of the relation between the biological
and the social differences between men and women
begins with two sets of facts about which there is no
dispute and which are laid out by Hubbard in Chapters
9, 10, and 11 of *The Politics of Women's Biology* and
in Lowe's essay in *Woman's Nature*. First, adult human
beings are, with few exceptions, divided into two types
that differ in their internal organs and in their external
genitalia, and these are associated with clearly different
roles in the reproductive process. There are, of course,
exceptional individuals with mixed or intermediate
anatomies, but the differences of all but a very few people are unambiguous. If we classify human beings by

these primary anatomical differences, into females and males, we find a large number of other anatomical and physiological characteristics that differ *on the average* between the sexes, but for which there is more or less variation between individuals of the same sex, and more or less overlap in range between the groups.

Both females and males secrete both estrogen and testosterone, although the relative amounts differ considerably. The amounts of the hormones change during development and vary with age, health, stress, exercise, and other aspects of experience. Breast development, skin texture, body hair, distribution of fat, size, and muscle mass all differ on the average between the anatomically defined sexes, but there are lots of smooth-skinned, fat, small, hairless, weak-muscled men, and many coarse, skinny, tall, hairy, and muscled women. All other claims about biological differences between the sexes, whether anatomical or psychological, are disputed and rest on weak evidence or no evidence at all.

The second set of indisputable facts, admirably summarized in Lila Leibowitz's very informative anthropological essay in *Woman's Nature*, are sociocultural generalizations. Every known human society has some division of labor by sex, although the particular tasks that are regarded as "men's work" and "women's work" vary considerably and may be reversed from one society to another. There are men's rites and women's rites, men's fashions and women's fashions, things that are forbidden to men and things that are forbidden to women, spheres of male power and of female power. Every society in every era has used the anatomical and reproductive dichotomy between male and female as a basis for a dichotomy in social organization along productive and ritual lines. For Hubbard and her colleagues, the question is: "What is the relation between the anatomical and social facts, and why do we care, anyway? Is anatomy destiny?"

There are roughly three positions one can take on the issues. The first is that some division of labor and rite is a structural property of human social organization, perhaps arising out of the very nature of human social manipulation of the world, and that sex difference, being the most obvious from birth and constantly in our consciousness as adults, is an arbitrary marker, neatly sorting people into two piles. Were the content of sex differences totally unrelated to the content of social differences, however, we would expect that the frequency with which institutional power, or property rights, or war making fell to males would be about the same as any of the three fell to females when we look over large numbers of human societies. While there are, indeed, matriarchal property rights and women warriors, these are far less frequent than their male equivalents, so the complete independence of the nature of anatomical differences from the content of social differences seems unlikely.

At the opposite extreme is the biological determinist view, now more fashionable than ever with the overestimation of the importance in human experience of DNA, that the sexual division of labor and power is the direct consequence of physiological and anatomical differences between men and women. The overt biological differences, in this view, are themselves causally efficacious, but more than that, they are signs of many other differences in brain structure and function which limit men and women in their roles. A large body of literature presses this claim, the most influential at present coming from sociobiologists. So E. O. Wilson writes of the sexual division of labor that

> The genetic bias is intense enough to cause a substantial division of labor even in the most free and most egalitarian of future societies. . . . Even with identical education and equal access to all professions, men are likely to continue to play a disproportionate role in political life, business and science.[6]

Nor do only male scientists make such claims. Camilla Benbow and her colleague Julian Stanley created a considerable flurry with their claim in 1980, published in *Science* and widely publicized in the press, that women are biologically inferior in mathematical reasoning so that, although they can indeed do humdrum mathematics, really creative work is beyond their limits. If women mathematicians are rare, it is because the brain structures of women, developing under the influence of too much estrogen and too little testosterone, simply cannot cope with Fermat's Last

6. E. O. Wilson, "Human Decency Is Animal," *The New York Times Magazine*, October 12, 1975, pp. 38–50.

Theorem. Wilson's and Benbow's claims remind one of Plato's assertion in the *Republic* that women have all the same qualitative abilities as men but in lesser degree (except pancake making).

But this determinist view of the division of labor and affect cannot be right either. Knitting and hand weaving, almost exclusively women's work now that they are outside the mainstream of production, were exclusively men's work 150 years ago. In like manner, there are now hundreds of women coal miners, just as mining is becoming more insecure and marginalized as a lifetime occupation. Of course, these examples can also be taken to prove that men run the world and preempt the occupations that count. But we knew that already. The question is why?

The third position, taken by Hubbard and her colleagues, demands that we distinguish the origin of social differentiation from the forces maintaining it. In this view, the division of reproductive labor, a direct consequence of the anatomical difference between the sexes, lies at the origin of social differences in work and social role. Under early conditions of production and in hunting-and-gathering societies, the producers and nurturers of children will be more sedentary and a division of labor, of group association, of spheres of power will develop.[7] The continued maintenance of labor and power differences, and their elaboration, however, depend on

---

7. For one version of this scenario, see Lila Leibowitz, "Origins of the Sexual Division of Labor," in *Woman's Nature*.

particular historical circumstances so that we are not bound to the aboriginal situation. Pregnancy and nursing, even in societies of low technological level, do not put an absolute constraint on women's labor. When intensive labor is required, as for example at harvest time in peasant agriculture, women are in the fields by necessity while pregnant or nursing. In technologically advanced societies with extremely low birth rates and high levels of technical support, again the relation between the reproductive and the sexual division of labor is broken.

While this theory of the sexual division of labor seems reasonable, it is not the version offered by textbooks of sociobiology, behavioral genetics, and so-called "bio-social anthropology." Of course, like all theories of the origin and maintenance of the sexual division of labor, it is a speculation, so we would not want to teach it to the unsophisticated mind as if it were objectively true. What is so seductive about biological explanations is that they seem to smell of material reality, even when they are equally speculative.

The biological determinist explanation of the inequalities between the sexes requires a program of research that will show the material basis for the different abilities and limitations of both men and women. But the asymmetry in status and power between the sexes results in an asymmetry of explanatory schemes. For most researchers, it is women who need to be explained, not men, who are, after all, the norm, just as it is homosexuals who need to be explained but not heterosexuals; there is no search for the "gene for heterosexuality." So women are described as victims of "raging hormones,"

regularly debilitated by menstruation, subject to irrational mood swings. As Hubbard observes, "No one has
suggested that men are just walking testicles, but again
and again women have been looked on as though they
were walking ovaries and wombs."

4.

The Politics of Women's Biology discusses the history
of claims for a biologically determined cognitive difference between the sexes. During the nineteenth century,
it was a common medical opinion that the brain and
the female reproductive organs were in competition for
energy, so that an educated woman would be a sterile
woman. Testicles, apparently, had their own sources of
energy. Women's brains have been claimed to be smaller
than men's, although actual measurement shows them
to be a bit larger in proportion to body size, and, anyway, no one has ever found a correlation between brain
size and human cognitive abilities. More recently, a variety of contradictory claims have been made of differences in brain structure between the sexes. In a famous
set of observations on people whose connection between
the right and left brain hemispheres had been severed,
Roger Sperry and his colleagues claimed a differentiation
in cognitive functions between the two halves. The left
hemisphere was said to determine logical thought,
verbal behavior, mathematical ability, and "executive"
decision, while the right hemisphere was said to con-

trol visual-spatial ability, emotion, and intuition. So the classical feminine qualities seemed to reside in the right side of the brain.

The problem is that on tests of verbal ability, women perform better than men, although men are better at visual-spatial tasks. It must be, then, according to an explanation favored by Sperry, that women are less "lateralized"—i.e., dependent on one side of the brain —than men, using both parts of their brains more or less equally. On the other hand, a competing hypothesis asserts that women are more lateralized than men and are right-brained, while men's abilities are more evenly spread across the two hemispheres. Neither party has attempted to *measure* the degree of lateralization in men and women, nor have they suggested how one could go about doing so. The stories are just stories.[8]

Indeed, much of the "evidence" for basic biological differences determining differential abilities and roles turns out to confuse observations with their causes and explanations. So E. O. Wilson reasons that "in hunter-gatherer societies, men hunt and women stay at home. This strong bias persists in most agricultural and industrial societies, and, on that ground alone, appears to have a genetic origin."[9]

Hubbard's major theme in *The Politics of Women's Biology* is that the modern conception of biological

---

8. See Susan Starr, "Sex Differences and the Dichotomization of the Brain," in *Genes and Gender*.

9. Wilson, "Human Decency Is Animal."

woman has been constructed by an ideologically motivated and largely, but not exclusively, male-dominated
science. That is, there is an intimate connection between the place of women in science and the science of
women's place. So long as biology as an enterprise is
almost exclusively a male occupation, a biased science,
masquerading as objective, will make unfounded claims
about women's biology that will justify the inferior status of women.

It is certainly the case that nearly all academic biologists, especially those with tenure, are men, despite the
fact that biology has been a popular subject for women
students. It is also clear to anyone who has spent his
life in the academy that departmental relations, both
formal and informal, resemble membership in a small
club, with all the exclusiveness and sense of uniqueness that is implied. It must be remembered that academics expend a major portion of their psychic energy
acting as gatekeepers to professional acceptance,
whether they are judging students, refereeing journal
articles and book manuscripts, or deciding upon who
may join their ranks. Thus academics confront the
contradictions of the meritocratic ideal in a particularly acute form. If they really are judging on merit,
why are there so few women in their ranks? Why do
women not have merit? Certainly they have had access
to education, even graduate education. So the fault
must be in their very natures as women. While this
may appear a pessimistic view of scientific objectivity,
even a tolerant review of what biologists have said
about women and the quality of the evidence and logic

that have been used makes it hard to come to any other conclusion.

One feminist reaction to the male bias of women's biology has been to attempt a female-biased biology, in which the female turns out to be the smarter sex, the gentler sex, the more humane sex, the sex that has a real feeling for nature. But, of course, the evidence for an innate female decency is as bad as that for genetically inbuilt male nastiness, and Hubbard will have none of it. On the surface, there seems to be a contradiction in her position. If, as she says, science is inevitably biased by the social, personal, and political positions of the people who do it, if science cannot hope for some neutral (or neuter) objectivity, then on what grounds can she criticize the received biology of woman as "bad" biology? Has she not fallen into the pit that intellectual conservatives have claimed lies at the feet of every relativist? If there are no empirically grounded values, only what the literary theorist Barbara Herrnstein Smith calls "contingencies of value," are there also nothing but contingencies of truth about the natural world?

Hubbard finesses this problem, as I do, by an appeal to very basic universal cultural agreements. We demand certain canons of evidence and argument that are formal and without reference to empirical content: a two-valued logic, in which every proposition must be true or false, but not both; the truth tables of Whitehead and Russell, which create the rules of reasoning for such propositions; the logic of statistical inference; the power of replicating experiments; the distinction between observations and causal claims. No natural scientist will deny these as necessary conditions of science, underlying all valid claims about the material world. But on these grounds alone, nearly all the biology of gender is bad science.

A second claim for a feminist science is that its metaphors, methodology, and worldview will necessarily differ from masculine science. The metaphors of science are, indeed, filled with the violence, voyeurism, and tumescence of male adolescent fantasy. Scientists "wrestle" with an always female nature, to "wrest from her the truth," or to "reveal her hidden secrets." They make "war" on diseases and "conquer" them. Good science is "hard" science; bad science (like that refuge of so many women, psychology) is "soft" science, and molecular biology, like physics, is characterized by "hard inference."[10] The method of science is largely reductionist, taking Descartes's clock metaphor as a basis for tearing the complex world into small bits and pieces to understand it, much as the archetypical small boy takes apart the real clock to see what makes it tick.

A feminist science would be, it is claimed, less reductionist, less ham-fisted, better able to understand organisms and ecological systems as functioning harmonious wholes. The material world is a world of relations among things, and women are said to be more concerned with

---

10. Although conscious of the flavor of these metaphors, we all use them. I note that in an article in *The New York Review*, "The Corpse in the Elevator" (January 20, 1983, pp. 34-37), I wrote of embryogenesis "yielding" to the Cartesian attack. But Hubbard herself "explodes" the gene myth.

the dynamic of relationships than are men. Hubbard will have none of this either. Because she rejects the innateness and "naturalness" of what are thought of as feminine characteristics, she rejects the claim that women as a class must alter the nature of the scientific process.

There is, however, no compelling logic here. There is no reason that the socially constructed "woman" could not, in fact, alter institutions as much as any innate one. To deny the innateness of the feminine is not to deny the potential power of the image and set of attitudes and behaviors that are characterized as feminine (although not restricted to women) to alter institutions in which a large number of women take part. So if it were really the case that women, by training and socialization, concentrated more on relationships between things than on the properties of the things themselves, they might reject the reductionist, Cartesian model of, say, the brain and study the central nervous system in a more interactionist, dialectical mode. This mode would place less emphasis on fixed functions of fixed localities in the brain, and see that organ at a more integrated level.[11] Were such a change in method successful in solving the outstanding problem of modern biology, the effect would be to radically alter the dominant mode of investigation and conceptualization in other branches of science as well. While such a future for science is conceivable, it is unlikely.

The problem is rather one of the nature of the

---

11. For an interactive view of the function and structure of the brain, see Jean-Pierre Changeux, *Neuronal Man: The Biology of Mind* (Pantheon, 1985) and my review of it in Chapter 3.

historical process. If women do, indeed, succeed in becoming a proportionate part of the community of scientists, they will do so slowly, against resistance, and each woman scientist, as a person creating a life for herself in a social institution, will almost surely take on the attitudes and behaviors of the great mass of its members, men. That is certainly the history of women scientists up until the present, who have been successful precisely in the degree to which they are indistinguishable in scientific method from their male colleagues.

Much has been made of a special quality that Barbara McClintock is said to have brought to biology, "a feeling for the organism,"[12] that is thought to be characteristic of women's science. Yet the early work on chromosome mechanics that brought McClintock fame and status in the scientific community (one of the few women ever elected to the National Academy of Sciences, she achieved that apotheosis forty years before her Nobel Prize and more recent public fame) was utterly reductionist and mechanist in the then-reigning tradition of cytology and genetics. Indeed, by a kind of social circularity, women who are to succeed as a minority in science will do so only if they are part of the methodological consensus and by their very success will strengthen that consensus. In all likelihood, science will capture women, not women science. So Hubbard may be quite wrong in principle, but right so far as historical

12. See Evelyn Fox Keller's biography of McClintock, *A Feeling for the Organism: The Life and Work of Barbara McClintock* (Freeman, 1983).

events are concerned when, in *The Politics of Women's Biology*, she "doubt[s] that women as gendered beings have something new or different to contribute to science."

Women may have nothing to contribute to science as "gendered beings," "but women as political beings do." That is, for Hubbard the consciousness of womanhood is the consciousness of oppression. That is what is meant by saying that the personal is political. When women enter science they do not do so to confront men with the feminine, but to confront a dominant class with its exclusive and oppressive attitudes and actions. And, in doing so, they make the institution of science better, because they force it to confront its own lack of objectivity, its failure to live up to its self-proclaimed canons.

The problem of science as she sees it is not that it embodies masculine as opposed to feminine values, but that it is a mirror of a structure of social domination, that it produces falsely "objective" legitimization of that structure, and in so doing fails to live up to its own standard. It is, after all, the established academy, including professors at Harvard, Stanford, and Princeton, that claimed authoritatively to have shown that blacks, Mediterraneans, and the working class in general were biologically inferior, using canons of evidence that violate even the rudimentary demands of logical and empirical demonstration.[13] As is so often the case, the most radical attack on an institution is the demand

---

13. For a documentation of the role played by major figures in academic science in creating a "scientific" racism, see Gould, *The Mismeasure of Man*, reviewed in Chapter 1, and Lewontin, Rose, and Kamin, *Not in Our Genes*.

that it live up to its own myth. It is not an attempt to overthrow it but an attempt to cleanse and perfect it. "Think not that I have come to destroy the law or the prophets. I have not come to destroy but to fulfill."

Yet the same assimilationist pressure that makes it unlikely that women will succeed in bringing uniquely feminine viewpoints into science makes it doubtful that women will have the desired anti-ideological effect that Hubbard hopes for. Women cannot be both outside and inside science. They come to science as outsiders, but in the process of entering, they become insiders, beneficiaries of the same social status as their male colleagues, with the same interest in legitimizing the status quo. After all, *they* made it, so why can't you?

The transformation of personal life that the outsider experiences when taken into the inside has powerful, although not inevitable, consequences for political views. The personal becomes political. And because human beings are the consequences not of internally fixed programs of the genes, but of a continuous psychic development within a social structure, personal histories may illuminate theoretical positions.

That is certainly the case for Ruth Hubbard. Her most personal book, *The Shape of Red: Insider/Outsider Reflections*, is a series of autobiographical accounts and, like all autobiographies, it reconstructs history to fulfill theory. Thus the personal becomes political not only in life but in our reconstructions of it. In broad outline, Hubbard's biographical facts are straightforward and not unfamiliar. As a child of the well-off professional middle class in Vienna, she was on the inside, but as a girl and a Jew in Vienna in the 1930s, she was on the outside. As an immigrant to America, she felt on the outside, but as a Radcliffe student, the child of reestablished professionals in Cambridge, she was on the inside. As a young woman in science, she was on the outside, but in marrying a gentile from Westchester, she felt she was back on the inside. Then after a divorce and remarriage with a Jewish professor at an oppressively WASPy and rather anti-Semitic Harvard, she was outside and inside.

Not a few readers of (and contributors to) *The New York Review* will recognize elements of their own lives. Personal history, in one sense, explains everything, yet it predicts nothing because the same life histories can be claimed to predict the subscription lists of both the *Monthly Review* and *Commentary*. Every biological object, but especially a human being, is the nexus of a large number of weakly acting causes. No one, or few, of those causes determines the life of the organism; so that what appear to be trivially different causal stories may have utterly different end products. It is this structure of interaction of multiple causal pathways that makes living creatures, even the scientist, free in a way that inanimate objects are not. That is why, in the end, biographies tell us so little yet exemplify so much about the complexities of development.

An Exchange

The exchange that follows was published in the July 14, 1994, issue of The New York Review.

SARAH BLAFFER HRDY, Professor of Anthropology at the University of California, Davis, writes:

Whether or not women are "superior" to men, is never a topic that I have found particularly meaningful. Hence, I was dismayed to read in an article by Richard Lewontin that "the feminist anthropologist Sarah Blaffer Hrdy thinks women [are superior because they] are naturally *more* crafty and acquisitive than men, and have been made so by evolution." I certainly don't think this, and more importantly don't understand how any fair-minded scholar could come to that conclusion based on anything that I have written.

In my 1981 book *The Woman That Never Evolved* (Harvard University Press), I noted that "widespread stereotypes devaluing the capacities and importance of women have not improved either their lot or that of human societies. But there is also little to be gained from countermyths that emphasize woman's natural innocence from lust for power, her cooperativeness and solidarity with other women . . ." (p. 190). I wrote this because I believed that competition between females for direct access to resources or to access to particular males who controlled resources was a more important selective force in primate evolution than had been recognized up to that point. This may be what Mr. Lewontin is referring to when he claims I think females are naturally "acquisitive." In 1974, I was the first to propose that female primates may mate with multiple males so as to confuse information available to males about paternity and thereby enhance the survival of subsequent offspring, since former consorts might be more disposed to help, or at least not to harm, possibly related offspring. This has been a controversial and influential idea, and perhaps this is why Mr. Lewontin attributed to me the notion that females are naturally "crafty" (though he omits the critical context for the emergence of that craftiness, namely a world where females are trying to hold their own in a system otherwise favoring male interests). Nevertheless, "crafty" and "acquisitive" are his words, not mine, and none of this has ever led me to conclude that females were "superior" (or inferior) to males. Rather, what I have written is that "sociobiology, if read as a prescription for life rather than a description of the way some creatures behave, makes it seem bad luck to be born either sex. . . ."

C. DAVISON ANKNEY, Professor of Zoology at the University of Western Ontario, London, Ontario, Canada, writes:

Richard Lewontin apparently spent so much time reading Ruth Hubbard's books that he's missed many recent papers about sex differences in brain size and about the relation between brain size and human cognitive abilities. Perhaps he was referring to himself when he stated: "As scientists grow older, they often give up research in favor of philosophy, history, or politics. . . . Scientific work creates that bank account of legitimacy which we can then spend on our political and humanist pursuits." He must be aware, however, that an overdrawn account can lead to bankruptcy.

Regardless, his claims that women's brains are proportionately larger than men's and that "no one has ever found a correlation between brain size and human cognitive abilities" are patently false. I recently published an analysis of autopsy data (in the 1992 issue of *Intelligence*) from 1,261 adults and showed, unequivocally, that after statistically controlling for differences in body size, men's brains average about 100 grams (8 percent) heavier than those of women. At my suggestion, Professor Philippe Rushton analyzed data from a stratified random sample of 6,325 US Army personnel and showed that after controlling for effects of age, stature, and body weight, the cranial capacity of men averaged 110 cm^3 larger than that of women. (This too was published in the 1992 issue of *Intelligence*.) Subsequently, Professor Nancy Andreasen used magnetic resonance imaging techniques that, in effect, create a three-dimensional model of the brain in vivo, and found a similar sex difference in brain size.

Since the turn of the century, numerous studies have shown that there is a positive correlation (+0.2) between various measures of head size and mental test scores in general intelligence as well as in spatial and reasoning ability. Recently, several independent studies

used magnetic resonance imaging to estimate brain volume in normal people and found an even higher positive correlation between brain volume and cognitive
abilities (+0.4, as reported by Professor Andreasen and
her colleagues in the 1993 issue of *American Journal
of Psychiatry*). The brain-size/intelligence relation has
been found independently in both men and women.

Women have proportionately smaller brains than
do men, but apparently have the same general intelligence test scores. Thus, I have proposed that the sex
difference in brain size relates to those intellectual
abilities at which men excel. Women excel in verbal
ability, perceptual speed, and motor coordination
within personal space; men do better on various spatial tests and on tests of mathematical reasoning. It
may require more brain tissue to process spatial information. Just as increasing word-processing power in
a computer may require extra capacity, increasing
three-dimensional processing, as in graphics, requires
a major jump in capacity. In support of this hypothesis
is the published observation (by Andreasen) that brain
size correlates most highly with performance tests in
men and with verbal tests in women.

Predictably, correlations between cognitive abilities
and overall brain size will be modest. First, much of
the brain is not involved in producing what we call
intelligence: variation in size/mass of that tissue will
reduce the correlation. Second, mental test scores, of
course, are not a perfect measure of intelligence and
thus, variation in such scores is not a perfect measure
of variation in intelligence. I suspect, however, that

not even Professor Lewontin would deny that human intelligence is directly related to brain function (I am unaware of evidence suggesting that intelligence is derived, for example, from the liver). The evidence is clear that human brain size is a measure, albeit imperfect, of brain function. Richard Lewontin, Ruth Hubbard, and others with "politically correct" agendas can ignore or even deny the existence of these fascinating aspects of human biology and behavior. They cannot, however, make them disappear.

RICHARD LEWONTIN replies:

Sarah Hrdy's direct complaint against me is just. Nowhere has she ever written that women are superior to men. Rather her point is that evolution by natural selection has made women who are "assertive, sexually active, or highly competitive, who adroitly manipulated male consorts, or who were as strongly motivated to gain high social status as they were to hold and carry babies" (*The Woman That Never Evolved*, p. 14). That is, women have been made by evolution into creatures that give them both equality with men in some ways and means of dominance over them (by "adroit manipulation") in others. Like other sociobiologists, she believes that human nature must be understood as having been molded effectively by natural selection to maximize the passage of genes, but she is concerned to correct what she sees as a sexist bias in most sociobiology that sees the operation of this natural selection as only on males. To be fair to her, she is also more circumspect than most of her sociobiological colleagues about how strong the evidence is for the story about natural selection and the hegemony of the genes in human affairs. Nevertheless, she obviously believes in the story enough to have written a book and a number of popular articles on the matter, based on the comparison between humans and apes. Moreover, while usually being careful to say that human beings are very flexible and therefore are not just like other animals, she sometimes slips back into a more simplistic sociobiological mode, as when she writes that in "species after species . . . primate males have been able to . . . translate superior fighting ability into *political* preeminence over the seemingly weaker and less competitive sex" (*The Woman That Never Evolved*, p. 16). Chimpanzee politics?

It must be said, however, that no sociobiologist has ever claimed that men are superior to women, *tout court*. The claim has been, rather, that men have built into them certain properties that give them contextual superiority over women in the same sense that a watch that keeps correct time is said to be superior to one that loses minutes and hours. It is not abstract or moral, but functional superiority that is at issue. Evolution has made men better able to do some things than women, and those are the things that make the world go round. Hrdy sees a balance of forces between the sexes, rather than an unconditional superiority of men, like the system of checks and balances in the United States Constitution. (I am indebted to Ruth Hubbard for pointing out this parallel.) Men threaten and women manipulate.

234 The innocent reader may be somewhat surprised by the snotty ad hominem tone of C. Davison Ankney's letter, a tone usually employed by injured authors whose books have been savaged in *The New York Review*. The mystery is solved by the revelation in Ankney's letter that "at my [Ankney's] suggestion, Professor Philippe Rushton analyzed data from a stratified random sample of 6,325 US Army personnel. . . ." This is not something that one would ordinarily admit in public, not to speak of deliberately calling attention to it in a widely read intellectual journal. What most of the readers may not know is that Professor Rushton has not confined himself to measuring heads. He attained a deservedly brief notoriety in the popular press, especially in Canada, for his interest in penises. He claimed that measurements of the length and angle of repose of that appendage in black and white men showed greater length and a more jaunty angle in blacks, which, according to him, agrees with blacks' well-known sexual aggressiveness. Ankney's self-conscious public alignment with the perpetrator of this kind of nineteenth-century silliness does not instill much confidence.

In fact, Ankney's published paper was not on any new data, but was an attempt at reanalysis, without access to the actual data, of a study by Ho et al.[14] which

---

14. K. C. Ho, U. Roessmann, J. V. Straumfjord, and G. Monroe, "Analysis of Brain Weight: Adult Brain Weight in Relation to Body Height, Weight, and Surface Area," *Archives of Pathology and Laboratory Medicine* 104 (1980), pp. 635–645.

had shown that after body-size adjustment there was no consistent difference in brain size between the sexes. The essential difference in the analysis rests on whether one should correct each subject's brain size for the subject's body size and then average these corrected values, in my and Ho et al.'s view the correct procedure, or to take ratios of group averages. This is hardly the forum for such a technical discussion, so let me simply restate in exact terms what Ho et al. claimed, a claim with which Ankney does not, in fact, disagree. If a person's brain size, say the weight of the brain, is divided by the person's body size, say the weight of the body, then the average value of these corrected brain sizes does not differ between the sexes. Depending on what measurement one uses to characterize size, either weight, surface area, or height, the results may show women with slightly larger or slightly smaller brains, but overall there is no consistent difference. It should be noted that since women and men have different average shapes and ratios of fat to bone and muscle, we really have no way of knowing what a "correct" method of accounting for body size would be.

Ankney also cites a study by Andreasen et al.[15] claiming that there is a small but real positive correlation between brain size and IQ scores. This is not the first such report, but reports and convincing demonstrations

15. N. C. Andreasen, M. Flaum, V. Swayze, D. S. O'Leary, R. Alliger, G. Cohen, J. Ehrhardt, and W. T. C. Yuh, "Intelligence and Brain Structure in Normal Individuals," *American Journal of Psychiatry*, Vol. 150, No. 1 (January 1993), pp. 130–134.

are two different things. The study in question used a small sample obtained by advertising for volunteers in a newspaper, who were then screened (by undisclosed methods) to get a sample of sixty-seven. Whatever went on in the process, an extraordinary product was created, because the test-subject sample had an average IQ score of 118, whereas the population at large has an average score of only 100 and only about one person in seven has an IQ score above 116. Conclusions from such a nonrepresentative study cannot be taken too seriously. As Andreasen et al. say, "Controversy has persisted for many years about whether there are significant relationships between size and function in the human brain." It still persists.

Finally, we need to observe a contradiction that Ankney also notices in his published writing. If women have smaller brains than men, and if smaller brains produce smaller IQ scores, then women should have lower IQ scores than men. But they don't. So what's up? Maybe women make better use of less brain mass, or maybe IQ tests have been biased in favor of women. These explanations call into question either male superior brain power or the objectivity of IQ tests, neither of which is congenial to people who think that there are important issues here that can be solved by weighing brains and giving IQ tests. Then, of course, there is always the possibility that there is nothing to explain, except how people come by their ideologies.

Chapter 7

Sex, Lies, and Social Science

"Sex, Lies, and Social Science" was first published in The New York Review of Books *of April 20, 1995, as a review of* Science in the Bedroom: A History of Sex Research, *by Vern L. Bullough (Basic Books, 1994);* The Social Organization of Sexuality: Sexual Practices in the United States, *by Edward Q. Laumann, John H. Gagnon, Robert T. Michael, and Stuart Michaels (University of Chicago Press, 1994); and* Sex in America: A Definitive Survey, *by Robert T. Michael, John H. Gagnon, Edward O. Laumann, and Gina Kolata (Little, Brown, 1994).*

I ONCE KNEW a man who was posted as a research scientist at an agricultural institute in what was then British Uganda. He told me with great frustration that he was having extreme difficulty in finding out whether his African assistants had actually carried out the procedures that he had prescribed because they had become so anxious to please their colonial bosses that they always answered "Yes" to every question asked. He claimed, however, that he had thought of a way around the problem. In the future he would always elicit the same information twice in such a way that the correct answer would be "Yes" the first time he asked and "No" the second. It apparently had not occurred to him that if his assistants really always answered "Yes" to every question, his scheme was doomed to failure.

My friend had discovered the fundamental methodological difficulty that faces every historian, biographer, psychotherapist, and reader of autobiography, the problem of self-report. How are we to know what is true if we must depend on what interested parties tell us? The historian and biographer, at least, have access to alternate sources and to the intersection of the independent stories of reporters with different axes to grind. We don't need Napoleon's *Mémoriale de Sainte-Hélène* or Wellington's papers to know who won at Waterloo, and neither source would have been enough for Hugo's description of it in Part Two of *Les Misérables*.

Public events have many private versions, but private events produce only a single public show. The readers of *The New York Review of Books* need only reread the January 12, 1995, issue to see the problem in two of its manifestations: one, in the autobiography of a scientist who has been engaged in contentious ideological battles over his scientific claims for half of his professional life,[1] and the other, in the bitter struggle over the reliability of repressed memories of childhood abuse.[2] A third, and even more difficult one, is the attempt to find out what people do in their quest for sexual gratification and why. The famous studies by Alfred Kinsey and his collaborators in the 1940s and 1950s which have become part of everyday reference as "The Kinsey Report," the later research by Masters and Johnson, and the more popularly read work of Shere Hite[3] are

---

1. E. O. Wilson, *Naturalist* (Island Press, 1994), reviewed by Jared Diamond, *The New York Review*, January 12, 1995, pp. 16-19.

2. "Victims of Memory: An Exchange," *The New York Review*, January 12, 1995, pp. 42-48. See also Frederick Crews's two-part article, "The Revenge of the Repressed," *The New York Review*, November 17 and December 1, 1994.

3. A. C. Kinsey, W. B. Pomeroy, and C. E. Martin, *Sexual Behavior in the Human Male* (Saunders, 1948); A. C. Kinsey, W. B. Pomeroy, C. E. Martin,

part of a long history of the science of "sexology." Vern Bullough's *Science in the Bedroom* is an extensive review of that unsatisfactory history. "Bedroom" is, of course, pure synecdoche, since no space that can contain one or more human beings appears to have been excluded from the possible sites of sex. The latest try at knowing who does what to whom, and how often, is the National Opinion Research Center's *The Social Organization of Sexuality*, completed just too late to be included in Bullough's historical survey. Suspecting that Americans would not be wholly indifferent to their findings, the research workers who produced *The Social Organization of Sexuality* also arranged with the well-regarded science journalist Gina Kolata to collaborate on a popular version, *Sex in America*, an *haute vulgarisation* of our *basses vulgarités*.

We all have created elaborate fictions, both conscious and unconscious, that we try to sell to ourselves and others as the real stories of our lives. The reader of conventional autobiography is, in principle at least, able to test some of the self-indulgences of autobiographers, since much of what is of general interest in a public life has been seen and heard by others who may be consulted. Moreover, autobiographers do not know from the beginning that they will publish a life story, so they may commit to writing, unthinkingly, rather contradictory

---

and P. H. Gebhard, *Sexual Behavior in the Human Female* (Saunders, 1953); W. H. Masters, and V.E. Johnson, *Human Sexual Response* (Little, Brown, 1966); S. Hite, *The Hite Report on Female Sexuality* (Knopf, 1979) and *The Hite Report on Male Sexuality* (Knopf, 1981).

material. But these provide only a theoretical possibility of looking for the truth, since, with not many exceptions, one must be a Napoleon before anyone will bother to check an author's memoirs against the record. For the most part, autobiography is a free ride into history. Repressed memories, too, are not entirely liberated from tests of their credibility. First, it may be that repressed memories simply do not exist so that every claim to them must be false. It might indeed be true, as claimed by Frederick Crews, that the entire experience of psychiatry and psychology speaks against the phenomenon. Second, even if repressed memories do, in fact, exist, and can be called to consciousness by appropriate techniques, the credibility of particular repressed memories is strained by their content. Sensible people can only scoff at reports of widespread Satanic rituals in which Babbitts consume the flesh and blood of babies.

There remains, however, one realm of self-report that seems utterly resistant to external verification. Given the social circumstances of sexual activity there seems no way to find out what people do "in the bedroom" except to ask them. But the answers they give cannot be put to the test of incredulity. Surely we believe that there is no sexual fantasy so outrageous and bizarre, no life of profligacy so exhausting, that it has not been realized by someone, somewhere, perhaps even by a reader of *The New York Review*'s personals. But if by someone, then why not by 17.4 percent of white males with two years of education beyond high school and with an annual income of $43,217? What behavior that is credible in individuals becomes incredible in the mass? The problem is to turn biography into science. If research produced by the National Opinion Research Center (NORC), the organization that epitomizes modern objective statistical social science, designed and analyzed by two distinguished service professors and a past president of the International Academy of Sex Research, carried out by a full-time project manager in charge of 220 interviewers, and resulting in a book of 718 pages, including 178 tables, 34 graphs, and 635 references, does not crack the problem of knowledge from self-report, then not just "sexology," but all of scientific sociology, is in deep trouble.

The motivations for the NORC study were two. First, given the evident importance of sex in people's lives, it is hard to see how there could be an adequate theory of social processes, not to speak of efficacious planning of social policy, without an understanding of the shape of people's sex lives. Unfortunately, previous social surveys of sex, as documented in *Science in the Bedroom*, were methodologically unsatisfactory. The flaws in these studies did not arise from a simple lack of technical sophistication. Bullough's tremendously informative analysis shows that sex surveys did not come out of a general demand by sociologists to document yet another central feature of social life, or from the desire of theoretical sociologists to provide empirical evidence for some overarching theory of social determination. Rather, they were an outgrowth of a variety of theories of the determination of individual sexuality, of ideological convictions about sex, and of a concern about sexual pathologies.

244 A major change took place between the end of the nineteenth century and the early years of the twentieth, as studies of sex ceased being a concern with pathologies and became part of a crusade for sexual liberation. The earlier tradition was represented by Krafft-Ebing's famous compilation of scores of case histories, *Psychopathia Sexualis*, regarded as a scandalously raw book by my parents, who could not refer to it except sotto voce, and who would have been indignant to know that we preadolescents still tittered over its discreet Latin descriptions. The new "sexology" was epitomized by Havelock Ellis, whose research was in the service of a universal appreciation of human sexuality in all its aspects, including its formerly taboo manifestations in masturbation and homosexuality. For Ellis, the term "abnormal" meant simply a deviation from the average, being descriptive rather than normative in its intent. And between Krafft-Ebing's ideal of sex and Ellis's realities falls the shadow of Freud, who began with pathology and wound up with the domestication of incestuous desires. The gathering of case histories by these and other students of sex, like Magnus Hirschfeld in his attempt to establish the normality of homosexuality, was an instrument of argument, a demonstration of perceived truths about sex.

As psychology and social theory became social science, so studies of sex took on more of the methodological apparatus of the natural sciences. What were once compilations of illustrative case studies became large samples in objective surveys with elaborate interview protocols and questionnaires that included information about other social variables, such as economic status. Yet these surveys remained in an ideological tradition. Bullough describes Kinsey as an objective scientist:

> His two major works, the male study in 1948 and the female study in 1952, serve as effective indicators of the change taking place in American society. Though Kinsey is known for his diligent interviewing and summation of data, his work is most significant because of his attempt to treat the study of sex as a scientific discipline, compiling and examining the data *and drawing conclusions from them without moralizing* [emphasis added].

But what Bullough has missed here is that discussing sex "without moralizing" is precisely the moral position that what people do with their erogenous zones is simply part of human natural history, that sex, in Ellis's sense, is normal, and that notions of abnormality and deviance can have only a statistical meaning. So Kinsey and his epigones do not represent a real break in the smooth history of sexology, which continues to reflect the changing social attitudes toward fun in bed.

Moreover, it seems clear that Kinsey and Masters and Johnson and Shere Hite knew what they would find in their surveys, namely that, putting aside the trivia of percentages, a substantial number of ordinary people will say that they do anything you care to name. The lack of statistical rigor in the sampling techniques of these earlier studies is a revelation not of technical sloppiness but of the studies as demonstrations of what their planners already believed they knew to be true. So the sexologists didn't really think it mattered how they got their samples, and it turns out that they were substantially right because, as I will argue, sampling technique is not the important issue.

The second reason that the NORC team thought that a new sex survey was needed was its relevance to the epidemiology of AIDS. Because AIDS is spread largely through certain sexual practices, an accurate estimate of the frequency of such practices, the way they are distributed through the population, and what the network of sexual partners looks like are all important variables in any model of the spread of the disease. If we are really interested in a useful epidemiological model of AIDS spread, not to speak of one that does not make the situation worse, we had better get the answers right. We have more than an academic interest in knowing whether self-report is a road to truth.

The National Health and Social Life Survey (NHSLS), to give the NORC study its full and revealing, or, rather, concealing title, was designed originally to respond to a federal request for proposals (RFP) issued by the National Institutes of Health on behalf of a coalition of federal agencies concerned with AIDS. The missing "S" word in the title of the survey was a deliberate reflection of the absence of any reference to sex in the ironically misleading title of the RFP, "Social and Behavioral Aspects of Fertility-Related Behavior." At the very least there is some anatomical confusion here. The attempt to mislead the prudes in the Bush administration did not work, however, and final approval of the project was never given. Nor did the change in administrations help, because the Democratic Congress explicitly prohibited the use of NIH funds for such a survey. In the end it was those fonts of immorality, the Robert Wood Johnson, the Rockefeller, Kaiser, Mellon, MacArthur, and Ford Foundations who came to the rescue. Freed from the constraint of asking only about AIDS-related sex, the survey could then really ask about "fertility-related behavior."

It is a characteristic of the design of scientific research that exquisite attention is devoted to methodological problems that can be solved, while the pretense is made that the ones that cannot be solved are really nothing to worry about. On the one hand, biologists will apply the most critical and demanding canons of evidence in the design of measuring instruments or in the procedure for taking an unbiased sample of organisms to be tested, but when asked whether the conditions in the laboratory are likely to be relevant to the situation in nature, they will provide a hand-waving intuitive argument filled with unsubstantiated guesses and prejudices because, in the end, that is all they can do. *The Social Organization of Sexuality* is a paradigm of the practice, made all the more objectionable by the air of methodological snootiness assumed by the authors when comparing their techniques with all the studies that have gone before. So they expend immense intellectual energy on the problem of taking a representative sample of Americans for an inquiry into their sex lives, but are rather cavalier about the question of whether people tell them the truth when asked.

The "sample survey" is the most highly developed technique of modern scientific sociology. Its purpose is to replace, by some objective measures, the impressionistic barroom wisdom of an older, more personal and reflective form of social commentary, in which an elaborate theory of social organization is built either on a prioris or on the commentator's necessarily limited autobiographical experience of what people are like. One can imagine *Leviathan* written not in 1651 but in 1951: "The condition of the English man between the ages of 16 and 55 with an income of less than £50 is a condition of war of 73.4% of everyone against 58.6% of everyone else."

A sample survey consists of two general procedures corresponding to the name of the process. First, it tries to characterize a population without examining every individual. That is, it is not a census of the entire population but an attempt to recover the same information that would appear in a total census from a small (usually very small) sample of the entire group. Even the efforts of the Bureau of the Census and the Internal Revenue Service to get hold of every one of us turn out, in practice, to produce only samples, and therein lies a serious problem. Are the people who are not included different in some systematic way from those who were caught? This issue has plagued the organizers of the decennial Census, who have been accused of under-counting the homeless, the aged, the young, the black, and the poor. And even if they did include everyone, it would not require the unstinting efforts of Pat Robertson to keep them from asking us all about oral sex. A sample survey begins with the assumption that one cannot ask everybody the questions of interest, and devotes considerable statistical sophistication to finding 3,432 people who will accurately represent 200 million postpubertal Americans.

Second, having decided whom to include in the sample, the survey must find a way of getting information. Sometimes, but remarkably infrequently, information can be acquired without the willing participation of the people sampled. Whether you like it or not, the state knows how much interest you earned in banks last year, and the number of cars per hour going across the George Washington Bridge on summer Sundays can be objectively determined. A good deal has been learned about patterns of consumption by measuring the output of garbage from urban households. But these are exceptions. For the most part social surveys depend on the answers people give on questionnaires, forms, and applications, or from other kinds of voluntary activities, and these are unreliable to different degrees.

One cannot know, for example, how many women suffer domestic assault by asking their husbands, or even count the ridership on the New York subways by the number of tokens taken in at the end of the day, although one of these estimates is clearly more reliable than the other. Nearly all the information that one would like to get about people is affected in some degree by the problem of self-report. The reader might try to imagine how he or she would get absolutely reliable information about the ages of living Americans independent of other social variables. Birth certificates do not tell who is still alive. Drivers' licenses only find people who drive, underrepresenting, like social security records, the urban poor.

There are, moreover, different depths of unreliability in the answers to different implied questions. If people lie when we ask their ages, they are misleading us about their actual ages, and they are revealing something, although we are not quite sure what, about their attitudes toward age. If they lie when we ask them about their attitudes, say, whether they dislike blacks, we will not only underestimate racism as a conscious prejudice but also fail to estimate accurately the amount of practiced discrimination. More subtly, the answer to the question contains information about attitudes toward attitudes, about whether people consider their prejudices to be shameful, yet we have no way of knowing how to disentangle this self-reflexive aspect of human consciousness.

The National Health and Social Life Survey's chief claim for its superiority over previous sex studies lies in its sampling methodology. The work of Kinsey, and of Masters and Johnson, were the efforts of "sexologists," investigators whose training and interest were not statistical but descriptive. It was sufficient for them that nonnegligible fractions of Americans engaged in a diversity of different practices. Kinsey, in particular, thought that picking people out of a hat would produce a sample of recalcitrant subjects who were unlikely to tell him what he wanted to know. Given Kinsey's liberatory ideology, it was not of the utmost importance to him whether his estimate of 10 percent for male homosexuality was accurate. It was true to life. Kinsey's samples made no pretense to be somehow numerically accurate representations of the entire population, but were what Edward O. Laumann and his colleagues in the NHSLS call "convenience samples," consisting of patients, friends, neighbors, relatives, employees, people who have answered ads soliciting subjects for an experiment, or who have filled in a questionnaire sent to them because they are on the list of a periodical or an organization.

In contrast, the NHSLS sample was a so-called "probability sample" meant to make precise the chance that any American would be included. The process occurred in two stages. First a "random sample" of nine thousand addresses drawn from the Census was taken so that every household in the nation was equally likely to be included. Of these, about 3,700 were useless because no one lived there, or were excluded because the household had no English speakers or anyone between the ages of eighteen and fifty-nine. The second stage was to increase the representation of black and Hispanic households by a known amount, producing a so-called "stratified sample," because it was felt that these groups would be insufficiently represented in the random address sample to get accurate statistics on them.

The very words "address sample" and "probability sample" seem to promise a technological process that is ideologically neutral and objective, yet the sampling process itself, meant to be the study's strongest point, is laden with social theory that is replicated and enlarged in the later analysis of the data. Social theory enters first at the moment of stratification. The claim that some groups need to be overrepresented in the sample is based on a prior theoretical commitment to the relevance of those group identifications as variables in the eventual analysis. If one believed that religion was likely to be an important variable in determining people's sexual behavior, then a study ought to include enough Buddhists, Confucians, Hindus, and Jews to see whether belief in Original Sin really matters. In view of the preponderance of Christians, it would be necessary to add in extra heathens, since there might not be any Hindus at all in a random sample of five thousand households. Some of the readers of *The New York Review* may be disappointed to learn that there were so few Jews in the NHSLS sample that nothing can be said about whether they get more kicks from vaginal intercourse, anal touching, oral sex, using vibrators, watching other people, or ten other categories of potentially stimulating practices.

But why choose religion a priori as a relevant variable except that there is a mass of conventional social theory that claims its importance? Laumann et al. say that although they wanted to study what they call "sexual scripts," the detailed network of relationships and practices, they could not because it was too hard. So,

> Much fine-grained cultural (and, to some extent, regional) variation in these scripts was beyond

our grasp. Similarly, our ability to measure people's networks was also quite limited. We could not ask them about ... their relationships with specific persons other than their sex partners.

Their solution was to sample and analyze sexual behavior according to a set of prior categories that is easy to define and accord with standard notions of social causation.

> Given these limitations, we adopted a primarily inductive approach using the types of information that are easier to collect accurately with large surveys, such as information about the respondent's gender, race (and ethnic background), age, education, marital status, and religious affiliation. Each of these characteristics or "statuses" *is a basic component of the self-identity of the individuals who possess them, organizes the patterning of social relationships, and organizes people's understanding of the social world around them* [emphasis added]. Of course, many other characteristics also possess these features; however, this basic set is both universally recognized and, in many cases, arguably most salient—hence the term *master statuses.*

"In many cases, arguably most salient"? So are these really the masters of our social and sexual lives or aren't they? This is not the last time in *The Social Organization of Sexuality* that the authors try to

254

finesse a deep question with a shallow phrase. As a matter of fact, these are not the only social variables that the survey asked about.

In a section of the questionnaire labeled with the ideologically neutral term "Demography," the survey asked detailed questions about what can only be described as "social class": Did your father or mother work for pay when you were fourteen? What did they actually do on the job? What kind of place did they work for? What was their education? How many hours a week do you work for pay? Describe in detail your job, your duties, what kind of place you work for, your wage rate.

Yet social class or anything like it is never discussed in the book, nor do these variables ever appear in the 178 tables and 34 graphs. Apparently it is not a "master status" variable. Indeed, reference of any kind to income appears only twice when we learn (Table 10.2) that "rich" people are slightly happier and much healthier than "poor" people, and that they are much more likely to be interested in sex and to succeed at it (Table 10.8). Perhaps poor people are just too tired out from trying to get through life. The authors make no comment. It might be claimed that the importance of the "master status" variables is justified after the fact by the results, since there are differences in reported sexual activity and practices among individuals falling in different groups. But aside from the obvious differences by sex, age, and marital status, the other "master variables" show surprisingly little variation in the answers given. For example, as a measure of promiscuity, one can ask what proportion of respondents report having had more than one sex partner in the last year. The answers were: 23.4 percent of males but only 11.7 percent of females, 32.2 percent of those aged 18–24 dropping to 18.5 percent at age 30–34, and 34.7 percent of never-marrieds but a mere 4.1 percent of those currently married. In contrast, there is hardly a difference between those who never finished high school and those with graduate school degrees (17.2 percent as opposed to 13.4 percent) and even less variation by professed religion: Jews (18.2 percent), fundamentalist Protestants (17.0 percent), Catholics (15.4 percent), or liberal Protestants (who at 15.0 percent seem the least "liberal" of all).

More important from a theoretical standpoint is the problem of proxy variables and the lack of independence between categories. To what extent are race and ethnicity, years of schooling, and even religion proxies for social class? To what extent are these variables themselves interrelated? Being black, poor, unemployed, and without a high school diploma go together, so which of these "master variables" is really the master, or are they all just ways of saying "lower class"?[4] Laumann et al. make some attempt to deal with the correlation between these variables when they analyze the causes of the stability of unions and the age at which

---

4. The term *Lumpenproletariat* has gone out of fashion, but a belief in its existence has come to mark neoconservative social theory, as exemplified by Murray and Herrnstein.

people enter them, but for the most part the categories are taken at their face value. From the very moment that a social survey sample is designed, the theoretical assumptions of the investigators about causal pathways in social determination come to dominate the study.

There is another peculiarity of the NHSLS sample that is particularly relevant because the study is said to be motivated by the need to make epidemiological models of the spread of AIDS. Because the sample is based on household address, the survey does not include the 3 percent of Americans (about 7.5 million) who do not live in households but are in institutions or are homeless. For many purposes ignoring 3 percent of the population is trivial, but for the epidemiology of AIDS it is precisely those in prison, in homeless shelters and on the streets, and in college dormitories who are most relevant. The prevalence of homosexual rape in prisons, the indiscriminate prostitution that characterizes drug addiction, and the relentless sexuality of college-age adolescents all mean that these ways of living are characterized by unusually complex networks of sexual contacts within the institutions, and with sexual practices that are likely to spread AIDS.

The authors of *Sexuality* take note of the exclusion of the institutionalized but pass it off by suggesting that "it would be wise to design and execute specialized research projects designed to study these groups separately." But studying these groups separately is precisely what it would not be wise to do. The prison, military, and college populations have a constant turnover, so that a very large fraction of the entire population has spent some protracted period as part of them. Perhaps the most important question that could have been asked of a male in the survey was "Have you ever spent time in prison?" If the answer were "yes," then the appropriate next question would not be about his sex practices in the last year but about what happened to him in Dannemora.

However, these were not regarded as "master status" variables. I would have thought that there is nothing like having been raped in prison to "organize people's understanding of the social world around them." The authors do not discuss it, and they may not even realize it, but mathematical and computer models of the spread of epidemics that take into account the real complexities of the problem often turn out, in their predictions, to be extremely sensitive to the quantitative values of the variables. Very small differences in variables can be the critical determinant of whether an epidemic dies out or spreads catastrophically, so the use of an inaccurate study in planning countermeasures can do more harm than does total ignorance.

The second major problem of sample surveys is to know what questions to ask and how to go about asking them. The founder of modern sociological research, William Fielding Ogburn, said that the central question for any claim of social theory was "How do you know it?"[5] The answer, alas, cannot be "Because

---

5. Quoted by A. J. Jaffe in his biographical article on Ogburn in the *International Encyclopedia of the Social Sciences*, Vol. 11 (MacMillan/Free Press, 1968), p. 277.

I asked." The problem for every sample survey is to know whether the answers are systematically untrue. Surveyed populations can lie in two ways. They can answer untruthfully, or they can fail to answer at all. This latter problem is known in the trade as "nonresponse bias." No matter how hard one tries, a significant portion of the sample that has been chosen will fail to respond, whether deliberately, through accident, lack of interest, or by force of circumstance.

It is almost always the case that those who do not respond are a nonrandom sample of those who are asked. Sometimes the problem is bad design. If you want to know how many women work outside the home you will not try to find out from a telephone survey that makes calls to people at home between nine AM and six PM. Much of the expertise of sample-survey designers is precisely in knowing how to avoid such mistakes. The real problem is what to do about people who deliberately avoid answering the very questions you want to ask. Are people who refuse to cooperate with sex surveys more prudish than others, and therefore more conservative than the population at large in their practices? Or are they more outrageous, yet sensitive to social disapprobation? Because they do not answer, and self-report is the only tool available, one can never know how serious the nonresponse bias may be. The best that can be done is to try to minimize the size of the nonresponding population by nagging, reasoning, and bribing. The NHSLS team tried all these approaches and finally got a response of 79 percent (3,432 households) after repeated visits, telephone calls, videotapes, and bribes ranging from $10 to an occasional $100. The result was that there were now three sample populations: those who were cooperative from the start, those who were reluctant but finally gave in, and those who refused to the end.

From an analysis of the eager and the reluctant it was concluded that for most questions there was no difference between the two, but that still leaves in the air the unanswerable question about the sex lives of those who found $100 an insufficient payment for their true confessions. If I can believe even half of what I read in *The Social Organization of Sexuality*, my own sex life is conventional to the point of being old-fashioned, and I wouldn't have cooperated for any price the NORC was likely to find in its budget.

Finally, we cannot avoid the main question, whether those who did respond, reluctantly or eagerly, told the truth. Far from avoiding the issue, the study team came back to this central question over and over, but their mode of answering it threatens the claim of sociology to be a science. At the outset they give the game away.

> *In the absence of any means to validate directly the data collected in a survey of sexual behavior,* these analyses assess data quality by checking for bias in the realized sample that might result from potential respondents' unwillingness to participate because of the subject matter, as well as by comparing results with other surveys. In every case, the results have greatly exceeded our expectations

> of what would be possible. They have gone a long
> way toward allaying our own concerns and skep-
> ticism.... [emphasis added].

In other words, people must be telling the truth because
other people have said it before and they say the same
thing even if reluctant to answer. That many people at
many times have independently claimed to have been
present at Satanic rituals or seen Our Lady descend at
Fatima, and that some of these witnesses have been
reluctant to testify at first, will presumably convince
Professor Laumann and his colleagues of the reality of
those events.

Again and again the problems of how we elicit the
truth when both conscious and unconscious distortions
may be suspected are dealt with disingenuously. Men
and women were interviewed by women and men
indiscriminately, and there was no attempt to match
race of the interviewer and race of the respondent.

> Will men and women respondents be affected in
> similar or different ways [by this mixing of sexes
> of interviewer and respondent]? Will people who
> have engaged in socially disapproved activities
> (e.g., same-gender sex, anal sex, prostitution, or
> extramarital sex relations) be equally likely to
> tell this to a male as to a female interviewer? At
> present, these questions remain unresolved empir-
> ically.... Although this issue is certainly impor-
> tant, ... we did not expect the effect of gender
> matching to be especially large or substantively

noteworthy. The experience and belief among
NORC survey research professionals was that the
quality of the interviewer was important but that
it was not necessarily linked to gender or race.

In other words, they don't know and hope the problem
will go away. While sex and race are "master status"
variables, "organizing the pattern of social relationships," apparently being interviewed about your sex
life is not part of social relationships. Instead of investigating the problem, the team "concentrated our time
and money on recruiting and training the best inter-
viewers we could find." That meant three days of a
"large-scale" training session in Chicago.

Anyway, why should anyone lie on a questionnaire
that was answered in a face-to-face interview with a
total stranger? After all, complete confidentiality was
observed. It is frightening to think that social science is
in the hands of professionals who are so deaf to human
nuance that they believe that people do not lie to themselves about the most freighted aspects of their own
lives, and that they have no interest in manipulating the
impression that strangers have of them. Only such
deafness can account for their acceptance, without the
academic equivalent of a snicker, of the result of a
NORC survey reporting that 45 percent of men between
the ages of eighty and eighty-four still have sex with a
partner.

It is not that the research team is totally unaware of
sensitivities. In addition to about a hundred face-to-
face interview questions, respondents were asked to fill

out four short printed forms that were placed by them in sealed "privacy" envelopes for later evaluation by someone other than the interviewer. Many of the questions were repetitions of questions asked in the personal interviews, following the common practice of checking on accuracy by asking the same question twice in different ways. Two matters were asked about, however, that were considered so jarring to the American psyche that the information was elicited only on the written forms: masturbation and total household income. Laumann et al. are not so deaf to American anxieties as it seemed.

There is, in fact, one way that the truth of the answers on a sex survey can be checked for internal consistency. A moment's reflection makes it clear that, discounting homosexual partners, the average number of sex partners reported by men must be equal to the average number reported by women. This is a variant on the economist Robert Solow's observation that the only law in economics is that the number of sales must be equal to the number of purchases. Yet, in the NHSLS study, and other studies like it, men report many more partners than women, roughly 75 percent more during the most recent five years of their lives. The reaction of the authors to this discrepancy is startling. They list "in no particular order" seven possible explanations, including that American men are having lots of sex out of the country, or that a few women are having hundreds of partners (prostitutes are probably underrepresented in an address sample, but prostitution was not regarded as a "master status" variable to be inquired about since presumably it is not a "basic concept of self-identity"). Our authors then say,

> We have not attempted to reconcile how much of the discrepancy that we observe can be explained by each of these seven logical possibilities, but we conjecture that the largest portion of the discrepancy rests with explanation 6.

Explanation 6 is that "either men may exaggerate or women may understate." So, in the single case where one can actually test the truth, the investigators themselves think it most likely that people are telling themselves and others enormous lies. If one takes the authors at their word, it would seem futile to take seriously the other results of the study. The report that 5.3 percent of conventional Protestants, 3.3 percent of fundamentalists, 2.8 percent of Catholics, and 10.7 percent of the nonreligious have ever had a same-sex partner may show the effect of religion on practice or it may be nothing but hypocrisy. What is billed as a study of "Sexual Practices in the United States" is, after all, a study of an indissoluble jumble of practices, attitudes, personal myths, and posturing.

The social scientist is in a difficult, if not impossible position. On the one hand, there is the temptation to see all of society as one's autobiography writ large, surely not the path to general truth. On the other, there is the attempt to be general and objective by pretending that one knows nothing about the experience of being human, forcing the investigator to pretend that people usually know and tell the truth about important issues, when we all know from our own lives how impossible that is. How, then, can there be a "social science"? The answer, surely, is to be less ambitious and stop trying to make sociology into a natural science although it is, indeed, the study of natural objects. There are some things in the world that we will never know and many that we will never know exactly. Each domain of phenomena has its characteristic grain of knowability. Biology is not physics, because organisms are such complex physical objects, and sociology is not biology, because human societies are made by self-conscious organisms. By pretending to a kind of knowledge that it cannot achieve, social science can only engender the scorn of natural scientists and the cynicism of humanists.

An Exchange

The exchange that follows was published in the May 25, 1995, issue of The New York Review.

EDWARD O. LAUMANN, JOHN H. GAGNON, ROBERT T. MICHAEL, and STUART MICHAELS, of the Department of Sociology at the University of Chicago, write:

We are puzzled by the review of our book, *The Social Organization of Sexuality*, because it is professionally incompetent and motivated by such an evident animus against the social sciences in general. We do not think it appropriate for a biologist, even a noted population geneticist whose empirical work is on the *Drosophila* fruit fly and other "simple" animals, to review a book that describes its principal task as formulating a social perspective on human sexual conduct in the United States. The notion that an economist, a sociologist, or a physicist should review professional work on population genetics would properly be greeted with derision. Lewontin's professional qualifications are of relevance in discussing his review since he himself asserts that his role as a *scientist* grants him the authority of special expertise for commenting on specific aspects of our book. Nowhere is that basis in expert knowledge evident in the innuendo and diatribe that constitute his review.

The central premise of Lewontin's review is that people routinely and pervasively lie about sexual behavior—indeed, it would seem all aspects of their lives—and thus none of the data from our survey of 3,432 people can be taken seriously. But Lewontin relates no systematic empirical information to substantiate his claim. Rather, he relies on a set of rhetorical devices that tendentiously advance his assertions.

Lewontin opens the review with an argument based on a false analogy. He discusses at length the problems of credibility in autobiographical statements and then asserts the analogical equivalence of autobiography and the self-reports given in response to our questions. The reader by now is supposed to be thinking, "I certainly would not tell anybody that I had sex with my spouse last night while clutching a yellow rubber ducky. I'd lie—at least about the rubber ducky." But autobiography, by definition, involves the public disclosure of the identity of the person. This sets in train all the motivations to create a favorable self-image in the minds of others and perhaps some of the outcomes Lewontin asserts. In contrast, we went to great lengths to guarantee the privacy, confidentiality, and anonymity of our respondents' answers as well as to provide a strong rationale for an individual to be candid and honest with us. We spent a great deal of time worrying about how we could check the reliability and honesty of our respondents' answers. While we readily admit that we were not always successful in securing full disclosure, his false analogy simply misses the point altogether.

Lewontin's next move is to provide an instance demonstrating the data's invalidity by discussing the large discrepancy between the average numbers of partners reported by men and women and the logical impossibility of such a situation assuming that they are recruiting their partners from a common pool. In the fifty-two-page chapter devoted to the numbers of sex partners, we explicitly discuss (on p. 174) the undesirability of using averages (means) to summarize the central tendencies of distributions as skewed and narrowly concentrated (with long, unevenly distributed tails) as these are. In addition, we explore in considerable detail the reasons for this discrepancy. Lewontin argues that if we could not get this "simple fact" right, it is evidence that all else is spurious. Error is a problem in all observations (including those in biology), how it is dealt with and its public recognition is the test of science. His decision to rest his case on this single issue without reference to its context forces us to conclude that he willfully misrepresented our analysis.

But he isn't satisfied with this. In an obscure footnote in the middle of the review that has no obvious relevance to our work at all, he mentions *The Bell Curve* by Herrnstein and Murray, the controversial book on racial differences in intelligence. Here we are being subjected to guilt by association. All readers of the *Review* surely know exactly what to think of these infamous social scientists. And we are insidiously being tossed into the pot with them for no other reason than we too are social scientists.

Finally, we have Lewontin's discussion of our finding that 45 percent of men between the ages of eighty and eighty-four claim to have sex partners. He chuckles at our credulity in reporting such patent nonsense, being just one more instance of our hopeless gullibility of believing everything we are told by our respondents. Now this is a rather nice instance of his tendentious and misleading use of our data to support his central claim that everybody is lying about their sex lives. The survey in question, the General Social Survey (GSS), is a widely known, high-quality, regularly conducted survey that professionally knowledgeable people rely on for estimating social trends of various sorts. It is sponsored by the National Science Foundation and has been subjected to regular scientific peer review for some twenty years. To the professional social scientist, it is well known to be a household-based sample that excludes the *institutionalized* parts of the population. Any number of census and other highly regarded survey studies have also noted that, due to differential mortality and other factors, older women are progressively more likely to be living alone. By age seventy, about 70 percent of women report, in the GSS, no sex partners in the past year. Older men, in contrast, are far more likely to be living with someone—the sex ratio is increasingly in their favor so far as the surplus of older women to older men is concerned. It is therefore not at all surprising that noninstitutionalized men in their eighties—presumably healthy enough to be living on their own—would have a fair chance of reporting that they have a sex partner. We discuss at length in the book the different meanings of sexuality across age, time, and social circumstance. We believe the answers are hardly likely to be crazed lies

by sex-starved octogenarians who are posturing like
teenagers for the edification of credulous social scientists.

The review is a pastiche of ill-informed personal
opinion that makes unfounded claims of relevant scientific authority and expertise. Readers of *The New
York Review of Books* deserve better.

RICHARD SENNETT, of New York University, writes:

In the course of Richard Lewontin's brilliant essay
"Sex, Lies, and Social Science" he remarks that if the
study he reviewed is typical of American scientific
sociology, then this discipline must be in "deep trouble." That's putting it mildly. American sociology has
become a refuge for the academically challenged.
Some universities have closed their sociology departments; many have decided the discipline merits little
new money.

Yet mere stupidity cannot explain the analytic
weaknesses of studies like the NORC sexuality project;
nor do social scientists so very gainfully employed in
such shops simply misunderstand the scientific enterprise. The difficulties with this research, like the larger
troubles of sociology, are political.

The British prime minister Margaret Thatcher
famously declared a generation ago, "There is no
society, only individuals and their families." In an
eerie way, much positivistic sociological research sub-
scribes to this antisocial nostrum. It does so, as in the
NORC study, by not probing subjects which resist

quantification; the usual disclaimer is that while such matters as the relation of sex and love may be important, they cannot be scientifically researched. Here is where politics enters; there's something comforting about sacrificing reality on the altar of research. The "dull science"—as Michel Foucault called American sociology—legitimates dissociation from the entanglements, contradictions, and difficulties of actual social experience. Dull knowledge has the same positive political value in Gingrich's America as it did in Thatcher's Britain. Lewontin complains of the superficiality of the NORC analysis, but maybe the very promise of a calming superficiality is what attracted so much money to this project.

However, if Lewontin's exposé is just, he uses a meat cleaver where a scalpel would have served him better. Is quantifying social phenomena an inherent evil, as at points in his essay he seems to suggest? Lewontin surely wouldn't deny that the Census Bureau provides useful and necessary information. In principle, survey research has its uses, in revealing how people think about themselves. (I found it both interesting and cheering that 45 percent of men between the ages of eighty and eighty-four in the NORC study reported still having sex with a partner, even if the aged have confused fantasy with fact.) Method per se isn't the issue.

I wish Lewontin had put his attack in a larger historical context. From its origins in Social Darwinism and the Progressive movement, American sociology has struggled with the contrary claims of those afflicted with physics envy and researchers—whether deploying numbers or words—more engaged in the dilemmas of society. In that struggle, Midwestern Protestant mandarins of positivist science often came into conflict with East Coast Jews who in turn wrestled with their own Marxist commitments; great quantitative researchers from abroad, like Paul Lazarsfeld at Columbia, sought to disrupt the complacency of native bean counters. In the last twenty years, more interesting "hard" sociological research has been done in medical, planning, and law schools, and better research on culture and society in the humanities departments, than in sociology departments. The *intellectual* enterprise of sociology is hardly represented by the dumbed-down study Lewontin rips apart.

What places like NORC command, like other reactionary enterprises, is money. To defend themselves, the minions of these institutions will undoubtedly attack Lewontin for being antiempirical, which will miss exactly his point, that their brand of science represses trenchant social evidence. My worry is that this repression is more than an academic evil. Sociology in its dumbed-down condition is emblematic of a society that doesn't want to know too much about itself.

RICHARD LEWONTIN replies:

It should come as no surprise to the readers of *The New York Review* that the authors of *The Social*

Organization of Sexuality did not like what I wrote. I confess to having amused myself over the last couple of weeks by imagining what their inevitable letter would contain. I was sure that they would challenge the competence of a biologist to judge social science, as indeed they have. I also imagined, and hoped, that they would raise a series of substantive objections to my characterization of their methodology, backed by various pieces of evidence of which the review took no account, so that we might engage in a revealing unpacking of the issues. In this, alas, I was too sanguine. Their letter makes no arguments, but relies on their disciplinary authority while repeating unsubstantiated and doubtful claims.

It is reasonable that Laumann et al. would have preferred to have their work reviewed by a member of their own school of sociology, someone sharing the same unexamined methodological assumptions. They could then avoid the always unpleasant necessity of justifying the epistemic basis on which the entire structure of their work depends. Their assertion of my incompetence, however, is off the mark. It is both temperamentally and ideologically repugnant to me to provide advertisements for myself, but as I do not want Laumann and his colleagues, or other readers of the *Review*, to avoid confronting the issues by a facile dismissal of my expertise, I am obliged to provide a CV. Although a biologist, I have a graduate degree in mathematical statistics and have taught the subject for forty years. About 10 percent of my technical publications, including a textbook of statistics, have been devoted to problems of statistical sampling, estimation, and hypothesis testing. More important, my biological work must be classified as methodological, my chief contribution to the field having been an analysis of the deep epistemological difficulties posed by the data of evolutionary genetics and the introduction of new experimental approaches specifically designed to overcome the ambiguities. Finally, my work on epistemological problems, produced both alone and together with philosophers of science, appears in standard philosophical journals.[6] Whatever may be at issue here, it is not competence.

Laumann et al. complain that the results of sample surveys were falsely analogized with autobiography. Either they do not understand the structure of analogical reasoning or, as is more likely, they were so annoyed by the review that they read it only impressionistically. No such analogy was drawn, nor was any argument from analogy made. On the contrary, autobiography, repressed memory, and survey interviews were given as three different examples of a general problem of deriving objective information from self-report. I drew a contrast between the possibility of verification in the first two cases and the virtual impossibility in the last. Here our authors touch on the central methodological issue. It is their view that, although people may lie or exaggerate in autobiographies because they are trying to create a public persona, they will tell the truth in

6. For a less technical and more generally accessible example, see "Facts and the Factitious in the Natural Sciences," *Critical Inquiry* 18 (1991), pp. 140–153.

anonymous interviews because there is no motivation to manipulate the impression that strangers have of us. Is it really true that quantitative sociologists are so divorced from introspection and so insensitive to social interactions that they take such a naive view of human behavior? Do they really believe all those things they hear from the person on the next bar stool or the seat next to them in the airplane? The Yellow Kid, who made a living from fleecing the gullible, used to say that anyone who could not con a banker ought to go into another line of work. Maybe, but before giving up, they should try professors of sociology. Putting aside subjective questions, haven't they even read the voluminous literature on the sociology of fashion? It is ironic that a student of "simple organisms" has to instruct those who inquire about human beings about the complexity of their objects of study.

First, Professor Laumann, people do not tell *themselves* the truth about their own lives. The need to create a satisfying narrative out of an inconsistent and often irrational and disappointing jumble of feelings and events leads each of us to write and rewrite our autobiographies inside our own heads, irrespective of whether anyone else is ever privy to the story. Second, these stories, which we then mistake for the truth, become the basis for further conscious manipulation and manufacture when we have exchanges with other human beings. If the investigators at NORC really do not care what strangers think of them, then they are possessed of an insouciance and hauteur otherwise unknown in Western society. It is precisely in the interaction with strangers who are not part of their social network, and who will never intersect their lives again, that people feel most free to embroider their life stories, because they will never be caught out.

Laumann et al. try to minimize the impact of the observed discrepancy in the number of sexual partners reported by men and by women. There is an attempt at obfuscation in a remark by Laumann and his colleagues about averages not containing as much information as more detailed frequency descriptions. True, but irrelevant, because in their data men consistently report more partners across the entire frequency distribution. Anyway, Laumann et al. do not deny the discrepancy. Indeed it is they who brought it up and discussed it in the book, and it is they, not I, who offered as the most likely explanation that men "exaggerate" and women "minimize" their sexual promiscuity. Then they try to discount the impact of the discrepancy on the study as a whole. After all, it is just one false note, and we cannot expect perfection. People may lie or fantasize about how many sexual partners they have, but we can take everything else they say at face value.

But this neatly ignores the fact that this comparison provides the *only* internal check on consistency that the study allows. I nowhere claimed that "all else is spurious," but rather that we are left in the unfortunate position of not knowing what is true when our only test fails. Then, in an extraordinary bit of academic chutzpah that turns the usual requirement for validation on its head, Laumann et al. say that it is up to those who question the data to demonstrate their unreliability. For years those of us who work on "simple" organisms have sheepishly accepted the burden of supporting our own claims, and the failure of the sole internal check on the validity of the data usually creates a certain difficulty in getting one's work published. *Autres pays, autres moeurs.*

I would not want to claim that we learn nothing from people's answers in sex surveys. One thing that they seem to establish is that individual fantasies follow cultural stereotypes. In the French equivalent of the NORC study involving over 20,000 telephone interviews, French men reported four times as many partners as French women![7] Of course, it may be that with the greater distance offered by the telephone, men feel freer to "exaggerate," but that explanation doesn't offer much solace to those who think that anonymity breeds truthfulness.

While Laumann and his colleagues believe that men exaggerate while they are aged between eighteen and fifty-nine, they (backed by the peer review panels of the National Science Foundation) seem to have complete confidence in the frankness of octogenarians. Perhaps, as men contemplate their impending mortality, the dread of something after death makes lying about sex seem risky. We must, however, at least consider the alternative that affirming one's continued sexual prowess in great age is a form of whistling in the dark.

---

7. A. Spira and N. Bajos, *Les comportements sexuels en France* (Paris: Documentation Française, 1993).

Far from having "an animus against the social sciences," I have considerable sympathy for the position in which sociologists find themselves. They are asking about the most complex and difficult phenomena in the most complex and recalcitrant organisms, without that liberty to manipulate their objects of study which is enjoyed by natural scientists. In comparison, the task of the molecular biologist is trivial. Living organisms are at the nexus of a large number of weakly determining causal pathways, and the classic method of studying such systems is to exaggerate the effect of one pathway while holding the others constant. When such experimental manipulation is not possible we have no recourse but to stand off and describe the system in all its complexity. The inevitable consequence is that the structure of inference is much looser and it becomes extremely difficult to test our explanations. How much worse is the situation of those observers whose objects of study have consciousness and who depend on the objects themselves to report on their own state.

The division between those who try to learn about the world by manipulating it and those who can only observe it has led, in natural science, to a struggle for legitimacy. The experimentalists look down on the observers as merely telling uncheckable just-so stories, while the observers scorn the experimentalists for their cheap victories over excessively simple phenomena. In biology the two camps are now generally segregated in separate academic departments where they can go about their business unhassled by the unbelievers. But the battle is unequal because the observers' consciousness of what it is to do "real" science has been formed in a world dominated by the manipulators of nature. The observers then pretend to an exactness that they cannot achieve and they attempt to objectify a part of nature that is completely accessible only with the aid of subjective tools.

Richard Sennett has formulated better, and with more authority than I could, the ideological issues in sociology. (Is he, too, incompetent?) He is, of course, right when he insists that quantitative information is important in sociology. Data on birth, death, immigration, marriage, divorce, social class, neighborhood, causes of mortality and morbidity, occupations, wage rates, and many other variables are indispensable for sociological investigations. My "meat cleaver" was never meant to sever those limbs from the body of knowledge. But it does not follow that collecting statistics, especially survey statistics with their utter ambiguity of interpretation, is sociology. A better model is Louis Chevalier's *Classes laborieuses et classes dangereuses.*[8] Chevalier's realization was that social phenomena could not be understood without the demographic statistics, but that these numbers can have no interpretation in themselves without a coherent narrative of social life. For contemporary life we have our own experience to help us understand the numbers. For the past we depend on literature, so the locales, characters, and events in the novels of Balzac, Hugo, and Sue form as much a part

---

8. Louis Chevalier, *Classes laborieuses et classes dangereuses à Paris pendant la première moitié du XIXème siècle* (Paris: Plon, 1958).

of the evidence about nineteenth-century Paris as the schedules of mortality and the tables of wage rates.

Even though the world is material and all its phenomena, including human consciousness, are products of material forces, we should not confuse the way the world is with our ability to know about it. Like it or not, there are a lot of questions that cannot be answered, and even more that cannot be answered exactly. There is nothing shameful in that admission.

Chapter 8

THE CONFUSION OVER CLONING

"The Confusion over Cloning" was first published *in* The New York Review of Books *of October 23, 1997, as a review of* Cloning Human Beings: Report and Recommendations of the National Bioethics Advisory Commission (Rockland, Maryland, June 1997).

THERE IS NOTHING like sex or violence for capturing the immediate attention of the state. Only a day after Franklin Roosevelt was told in October 1939 that both German and American scientists could probably make an atom bomb, a small group met at the President's direction to talk about the problem and within ten days a committee was undertaking a full-scale investigation of the possibility. Just a day after the public announcement on February 23, 1997, that a sheep, genetically identical to another sheep, had been produced by cloning, Bill Clinton formally requested that the National Bioethics Advisory Commission "undertake a thorough review of the legal and ethical issues associated with the use of this technology. . . ."

The President had announced his intention to create an advisory group on bioethics eighteen months before, on the day that he received the disturbing report of the cavalier way in which ionizing radiation had been administered experimentally to unsuspecting subjects.[1] The commission was finally formed, after a ten-month delay, with Harold Shapiro, President of Princeton, as chair and a membership consisting largely of academics from the fields of philosophy, medicine, public health, and law, a representation from government and private foundations, and the chief business officer of a pharmaceutical company. In his letter to the commission the President referred to "serious ethical questions, particularly with respect to the possible use of this technology to clone human embryos" and asked for a report within ninety days. The commission missed its deadline by only two weeks.

In order not to allow a Democratic administration sole credit for grappling with the preeminent ethical issue of the day, the Senate held a day-long inquiry on March 12, a mere three weeks after the announcement of Dolly. Lacking a body responsible for any moral issues outside the hanky-panky of its own membership, the Senate assigned the work to the Subcommittee on Public Health and Safety of the Committee on Labor and Human Resources, perhaps on the grounds that cloning is a form of the production of human resources. The testimony before the subcommittee was concerned not with issues of the health and safety of labor but with the same ethical and moral concerns that preoccupied the bioethics commission. The witnesses representing the biotechnology industry were especially careful to assure the senators that they would not

1. Report of the specially created Advisory Committee on Human Radiation Experiments (October 3, 1995).

dream of making whole babies and were interested in cloning solely as a laboratory method for producing cells and tissues that could be used in transplantation therapies.

It seems pretty obvious why, just after the Germans' instant success in Poland, Roosevelt was in a hurry. The problem, as he said to Alexander Sachs, who first informed him about the possibility of the Bomb, was to "see that the Nazis don't blow us up." The origin of Mr. Clinton's sense of urgency is not so clear. After all, it is not as if human genetic clones don't appear every day of the week, about thirty a day in the United States alone, given that there are about four million births a year with a frequency of identical twins of roughly 1 in 400.[2] So it cannot be the mere existence of doppelgängers that creates urgent problems (although I will argue that parents of twins are often guilty of a kind of psychic child abuse). And why ask the commission on bioethics rather than a technical committee of the

---

2. In fact, identical twins are genetically *more* identical than a cloned organism is to its donor. All the biologically inherited information is not carried in the genes of a cell's nucleus. A very small number of genes, sixty out of a total of 100,000 or so, are carried by intracellular bodies, the mitochondria. These mitochondrial genes specify certain essential enzyme proteins, and defects in these genes can lead to a variety of disorders. The importance of this point for cloning is that the egg cell that has had its nucleus removed to make way for the genes of the donor cell has not had its mitochondria removed. The result of the cell fusion that will give rise to the cloned embryo is then a mixture of mitochondrial genes from the donor and the recipient. Thus, it is not, strictly speaking, a perfect genetic clone of the donor organism. Identical twins, however, *are* the result of the splitting of a fertilized egg and have the same mitochondria as well as the same nucleus.

National Institutes of Health or the National Research Council? Questions of individual autonomy and responsibility for one's own actions, of the degree to which the state ought to interpose itself in matters of personal decision, are all central to the struggle over smoking, yet the bioethics commission has not been asked to look into the bioethics of tobacco, a matter that would certainly be included in its original purpose.

The answer is that the possibility of human cloning has produced a nearly universal anxiety over the consequences of hubris. The testimony before the bioethics commission speaks over and over of the consequences of "playing God." We have no responsibility for the chance birth of genetically identical individuals, but their deliberate manufacture puts us in the Creation business, which, like extravagant sex, is both seductive and frightening. Even Jehovah botched the job despite the considerable knowledge of biology that He must have possessed, and we have suffered the catastrophic consequences ever since. According to Haggadic legend, the Celestial Cloner put a great deal of thought into technique. In deciding on which of Adam's organs to use for Eve, He had the problem of finding tissue that was what the biologist calls "totipotent," that is, not already committed in development to a particular function. So He cloned Eve

> not from the head, lest she carry her head high in arrogant pride, not from the eye, lest she be wanton-eyed, not from the ear lest she be an eavesdropper, not from the neck lest she be insolent,

not from the mouth lest she be a tattler, not from
the heart lest she be inclined to envy, not from the
hand lest she be a meddler, not from the foot lest
she be a gadabout

but from the rib, a "chaste portion of the body." In
spite of all the care and knowledge, something went
wrong, and we have been earning a living by the sweat
of our brows ever since. Even in the unbeliever, who
has no fear of sacrilege, the myth of the uncontrollable
power of creation has a resonance that gives us all
pause. It is impossible to understand the incoherent
and unpersuasive document produced by the National
Bioethics Advisory Commission except as an attempt
to rationalize a deep cultural prejudice, but it is also
impossible to understand it without taking account of
the pervasive error that confuses the genetic state of an
organism with its total physical and psychic nature as a
human being.

After an introductory chapter placing the issue of
cloning in a general historical and social perspective,
the commission begins with an exposition of the technical details of cloning and with speculations on the reproductive, medical, and commercial applications that are
likely to be found for the technique. Some of these
applications involve the clonal reproduction of genetically engineered laboratory animals for research or the
wholesale propagation of commercially desirable livestock; but these raised no ethical issues for the commission, which, wisely, avoided questions of animal rights.

Specifically human ethical questions are raised by two possible applications of cloning. First, there are circumstances in which parents may want to use techniques of assisted reproduction to produce children with a known genetic makeup for reasons of sentiment or vanity or to serve practical ends. Second, there is the possibility of producing embryos of known genetic constitution whose cells and tissues will be useful for therapeutic purposes. Putting aside, for consideration in a separate chapter, religious claims that human cloning violates various scriptural and doctrinal prescriptions about the correct relation between God and man, men and women, husbands and wives, parents and children, or sex and reproduction, the commission then lists four ethical issues to be considered: individuality and autonomy, family integrity, treating children as objects, and safety.

The most striking confusion in the report is in the discussion of individuality and autonomy. Both the commission report and witnesses before the Senate subcommittee were at pains to point out that identical genes do not make identical people. The fallacy of genetic determinism is to suppose that the genes "make" the organism. It is a basic principle of developmental biology that organisms undergo a continuous development from conception to death, a development that is the unique consequence of the interaction of the genes in their cells, the temporal sequence of environments through which the organisms pass, and random cellular processes that determine the life, death, and transformations of cells. As a result, even the fingerprints of identical twins are not identical. Their temperaments, mental processes, abilities, life choices, disease histories, and deaths certainly differ despite the determined efforts of many parents to enforce as great a similarity as possible.

Frequently twins are given names with the same initial letter, dressed identically with identical hair arrangements, and given the same books, toys, and training. There are twin conventions at which prizes are offered for the most similar pairs. While identical genes do indeed contribute to a similarity between them, it is the pathological compulsion of their parents to create an inhuman identity between them that is most threatening to the individuality of genetically identical individuals.

But even the most extreme efforts to turn genetic clones into human clones fail. As a child I could not go to the movies or look at a picture magazine without being confronted by the genetically identical Dionne quintuplets, identically dressed and coiffed, on display in "Quintland" by Dr. Dafoe and the Province of Ontario for the amusement of tourists. This enforced homogenization continued through their adolescence, when they were returned to their parents' custody. Yet each of their unhappy adulthoods was unhappy in its own way, and they seemed no more alike in career or health than we might expect from five girls of the same age brought up in a rural working-class French Canadian family. Three married and had families. Two trained as nurses, two went to college. Three were attracted to a religious vocation, but only one made it a career. One died in a convent at age twenty, suffering from epilepsy, one at age thirty-six, and three remain alive at sixty-three. So much for the doppelgänger phenomenon. The notion of "cloning Einstein" is a biological absurdity.

The Bioethics Advisory Commission is well aware of the error of genetic determinism, and the report devotes several pages to a sensible and nuanced discussion of the difference between genetic and personal identity. Yet it continues to insist on the question of whether cloning violates an individual human being's "unique qualitative identity."

> And even if it is a mistake to believe such crude genetic determinism according to which one's genes determine one's fate, what is important for one-self is whether one *thinks* one's future is open and undetermined, and so still to be largely determined by one's own choices [p. A8, emphasis added].

Moreover, the problem of self-perception may be worse for a person cloned from an adult than it is for identical twins, because the already fully formed and defined adult presents an irresistible persistent model for the developing child. Certainly for the general public the belief is widely expressed that a unique problem of identity is raised by cloning that is not already present for twins. The question posed by the commission, then, is not whether genetic identity per se destroys individuality, but whether the erroneous state of public understanding of biology will undermine an individual's own sense of uniqueness and autonomy.

Of course it will, but surely the commission has chosen the wrong target of concern. If the widespread genomania propagated by the press and by vulgarizers of science produces a false understanding of the dominance that genes have over our lives, then the appropriate response of the state is not to ban cloning but to engage in a serious educational campaign to correct the misunderstanding. It is not Dr. Wilmut and Dolly who are a threat to our sense of uniqueness and autonomy, but popularizers like Richard Dawkins who describes us as "gigantic lumbering robots" under the control of our genes that have "created us, body and mind."

Much of the motivation for cloning imagined by the commission rests on the same mistaken synecdoche that substitutes "gene" for "person." In one scenario a self-infatuated parent wants to reproduce his perfection or a single woman wants to exclude any other contribution to her offspring. In another, morally more appealing, story a family suffers an accident that kills the father and leaves an only child on the point of death. The mother, wishing to have a child who is the biological offspring of her dead husband, uses cells from the dying infant to clone a baby. Or what about the sterile man whose entire family has been exterminated in Auschwitz and who wishes to prevent the extinction of his genetic patrimony?

Creating variants of these scenarios is a philosopher's parlor game. All such stories appeal to the same impetus that drives adopted children to search for their "real," i.e., biological, parents in order to discover their own "real" identity. They are modern continuations of an earlier preoccupation with blood as the carrier of an individual's essence and as the mark of legitimacy. It is not the possibility of producing a human being with a copy of someone else's genes that has created the difficulty or that adds a unique element to it. It is the fetishism of "blood" which, once accepted, generates an immense array of apparent moral and ethical problems. Were it not for the belief in blood as essence, much of the motivation for the cloning of humans would disappear.

The cultural pressure to preserve a biological continuity as the form of immortality and family identity is certainly not a human universal. For the Romans, as for the Japanese, the preservation of family interest was the preeminent value, and adoption was a satisfactory substitute for reproduction. Indeed, in Rome the foster child (*alumnus*) was the object of special affection by virtue of having been adopted, i.e., acquired by an act of choice.

The second ethical problem cited by the commission, family integrity, is neither unique to cloning nor does it appear in its most extreme form under those circumstances. The contradictory meanings of "parenthood" were already made manifest by adoption and the old-fashioned form of reproductive technology, artificial insemination from anonymous semen donors. Newer technology like in vitro fertilization and implantation of embryos into surrogate mothers has already raised issues to which the possibility of cloning adds nothing. A witness before the Senate subcommittee suggested that the "replication of a human by cloning would radically alter the definition of a human being by producing the world's first human with a single genetic parent."[3] Putting aside the possible priority of the case documented in Matthew 1:23, there is a confusion here. A child by cloning has a full double set of chromosomes like anyone else, half of which were derived from a mother and half from a father. It happens that these chromosomes were passed through another individual, the cloning donor, on their way to the child. That donor is certainly not the child's "parent" in any biological sense, but simply an earlier offspring of the original parents. Of course this sibling may *claim* parenthood over its delayed twin, but it is not obvious what juridical or ethical principle would impel a court or anyone else to recognize that claim.

There is one circumstance, considered by the commission, in which cloning is a biologically realistic solution to a human agony. Suppose that a child, dying of leukemia, could be saved by a bone marrow replacement. Such transplants are always risky because of immune incompatibilities between the recipient and the donor, and these incompatibilities are a direct consequence of genetic differences. The solution that presents itself is to use bone marrow from a second, genetically identical, child who has been produced by cloning from the first.[4] The risk to a bone marrow donor is not great, but suppose it were a kidney that

---

3. G. J. Annas, "Scientific discoveries and cloning: Challenges for public policy," testimony of March 12, 1997.

4. There is always the possibility, of course, that gene mutations have predisposed the child to leukemia, in which case the transplant from a genetic clone only propagates the defect.

was needed. There is, moreover, the possibility that the fetus itself is to be sacrificed in order to provide tissue for therapeutic purposes. This scenario presents in its starkest form the third ethical issue of concern to the commission, the objectification of human beings. In the words of the commission:

> To objectify a person is to act towards the person without regard for his or her own desires or well-being, as a thing to be valued according to externally imposed standards, and to control the person rather than to engage her or him in a mutually respectful relationship.

We would all agree that it is morally repugnant to use human beings as mere instruments of our deliberate ends. Or would we? That's what I do when I call in the plumber. The very words "employment" and "employee" are descriptions of an objectified relationship in which human beings are "thing(s) to be valued according to externally imposed standards." None of us escapes the objectification of humans that arises in economic life. Why has no National Commission on Ethics been called into emergency action to discuss the conceptualization of human beings as "factory hands" or "human capital" or "operatives"? The report of the Bioethics Advisory Commission fails to explain how cloning would significantly increase the already immense number of children whose conception and upbringing were intended to make them instruments of their parents' frustrated ambitions, psychic fantasies, desires for immortality, or property calculations.

Nor is there a simple relation between those motivations and the resulting family relations. I myself was conceived out of my father's desire for a male heir, and my mother, not much interested in maternity, was greatly relieved when her first and only child filled the bill. Yet, in retrospect, I am glad they were my parents. To pronounce a ban on human cloning because sometimes it will be used for instrumental purposes misses both the complexity of human motivation and the unpredictability of developing personal relationships. Moreover, cloning does not stand out from other forms of reproductive technology in the degree to which it is an instrument of parental fulfillment. The problem of objectification permeates social relations. By loading all the weight of that sin on the head of one cloned lamb, we neatly avoid considering our own more general responsibility.

The serious ethical problems raised by the prospect of human cloning lie in the fourth domain considered by the bioethics commission, that of safety. Apparently, these problems arise because cloned embryos may not have a proper set of chromosomes. Normally, a sexually reproduced organism contains in all its cells two sets of chromosomes, one received from its mother through the egg and one from the father through the sperm. Each of these sets contains a complete set of the different kinds of genes necessary for normal development and adult function. Even though each set has a complete repertoire of genes, for reasons that are not well understood we must have two sets and only two sets to complete normal development. If one of the chromosomes should accidentally be present in only one copy or in three, development will be severely impaired.

Usually we have exactly two copies in our cells because in the formation of the egg and sperm that combined to produce us, a special form of cell division occurs that puts one and only one copy of each chromosome into each egg and each sperm. Occasionally, however, especially in people in their later reproductive years, this mechanism is faulty and a sperm or egg is produced in which one or another chromosome is absent or present more than once. An embryo conceived from such a faulty gamete will have a missing or extra chromosome. Down's syndrome, for example, results from an extra Chromosome 21, and Edward's syndrome, almost always lethal in the first few weeks of life, is produced by an extra Chromosome 18.

After an egg is fertilized in the usual course of events by a sperm, cell division begins to produce an embryo, and the chromosomes, which were in a resting state in the original sperm and egg, are induced to replicate new copies by signals from the complex machinery of cell division. The division of the cells and the replication of more chromosome copies are in perfect synchrony so every new cell gets a complete exact set of chromosomes just like the fertilized egg. When clonal reproduction is performed, however, the events are quite different. The nucleus containing the egg's chromosomes are removed and the egg cell is fused with a cell containing a nucleus from the donor that already contains a full duplicate set of chromosomes. These chromosomes are not necessarily in the resting state and so they may divide out of synchrony with the embryonic cells. The result will be extra and missing chromosomes so that the embryo will be abnormal and will usually, but not necessarily, die.

The whole trick of successful cloning is to make sure that the chromosomes of the donor are in the right state. However, no one knows how to make sure. Dr. Wilmut and his colleagues know the trick in principle, but they produced only one successful Dolly out of 277 tries. The other 276 embryos died at various stages of development. It seems pretty obvious that the reason the Scottish laboratory did not announce the existence of Dolly until she was a full-grown adult sheep is that they were worried that her postnatal development would go awry. Of course, the technique will get better, but people are not sheep and there is no way to make cloning work reliably in people except to experiment on people. Sheep were chosen by the Scottish group because they had turned out in earlier work to be unusually favorable animals for growing fetuses cloned from embryonic cells. Cows had been tried but without success. Even if the methods could be made eventually to work as well in humans as in sheep, how many human embryos are to be sacrificed, and at what stage of their development?[5] Ninety percent of the loss of the experimental sheep embryos was at the so-called

---

5. It has recently been announced that success in cloning in cows is almost at hand, but by an indirect method that, if applied in humans, raises the following ethical problem. The method involves cloning embryos from adult cells,

"morula" stage, hardly more than a ball of cells. Of the twenty-nine embryos implanted in maternal uteruses, only one showed up as a fetus after fifty days in utero, and that lamb was finally born as Dolly.

Suppose we have a high success rate of bringing cloned human embryos to term. What kinds of developmental abnormalities would be acceptable? Acceptable to whom? Once again, the moral problems said to be raised by cloning are not unique to that technology. Every form of reproductive technology raises issues of lives worth living, of the stage at which an embryo is thought of as human, as having rights including the juridical right to state protection. Even that most benign and widespread prenatal intervention, amniocentesis, has a nonnegligible risk of damaging the fetus. By concentrating on the acceptability of cloning, the commission again tried to finesse the much wider issues.

They may have done so, however, at the peril of legitimating questions about abortion and reproductive technology that the state has tried to avoid, questions raised from a religious standpoint. Despite the secular basis of the American polity, religious forces have over and over played an important role in influencing state policy. Churches and religious institutions were leading actors in the abolitionist movement and the Under-

---

but then breaking up the embryos to use their cells for a second round of cloning. No cloned calf was yet born as of August 1, 1997, but ten are well established *en ventre leurs mères*. Even if they all reach an unimpaired adulthood, they will owe their lives to many destroyed embryos.

ground Railroad,[6] the modern civil rights movement and the resistance to the war in Vietnam. In these instances religious forces were part of, and in the case of the civil rights movement leaders of, wider social movements intervening on the side of the oppressed against then-reigning state policy. They were both liberatory and representative of a widespread sentiment that did not ultimately depend upon religious claims.

The present movements of religious forces to intervene in issues of sex, family structure, reproductive behavior, and abortion are of a different character. They are perceived by many people, both secular and religious, not as liberatory but as restrictive, not as intervening on the side of the wretched of the earth but as themselves oppressive of the widespread desire for individual autonomy. They seem to threaten the stable accommodation between Church and State that has characterized American social history. The structure of the commission's report reflects this current tension in the formation of public policy. There are two separate chapters on the moral debate, one labeled "Ethical Considerations" and the other "Religious Perspectives." By giving a separate and identifiable voice to explicitly religious views the commission has legitimated religious conviction as a front on which the issues of sex, reproduction, the definition of the family, and the status of fertilized eggs and fetuses are to be fought.

---

6. An example was the resistance to the Fugitive Slave Acts by the pious Presbyterians of Oberlin, Ohio, an excellent account of which may be found in Nat Brandt, *The Town That Started the Civil War* (Syracuse University Press, 1990).

The distinction made by the commission between "religious *perspectives*" and "ethical *considerations*" is precisely the distinction between theological hermeneutics—interpretation of sacred texts—and philosophical inquiry. The religious problem is to recognize God's truth. If a natural family were defined as one man, one woman, and such children as they have produced through loving procreation; if a human life, imbued by God with a soul, is definitively initiated at conception; if sex, love, and the begetting of children are by revelation morally inseparable; then the work of bioethics commissions becomes a great deal easier. Of course, the theologians who testified were not in agreement with each other on the relevant matters, in part because they depend on different sources of revelation and in part because the meaning of those sources is not unambiguous. So some theologians, including Roman Catholics, took human beings to be "stewards" of a fixed creation, gardeners tending what has already been planted. Others, notably Jewish and Islamic scholars, emphasized a "partnership" with God that includes improving on creation. One Islamic authority thought that there was a positive imperative to intervene in the works of nature, including early embryonic development, for the sake of health.

Some Protestant commentators saw humans as "co-creators" with God and so certainly not barred from improving on present nature. In the end, some religious scholars thought cloning was definitively to be prohibited, while others thought it could be justified under some circumstances. As far as one can tell, fundamentalist Protestants were not consulted, an omission that rather weakens the usefulness of the proceedings for setting public policy. The failure to engage directly the politically most active and powerful American religious constituency, while soliciting opinions from a much safer group of "religious scholars," can only be understood as a tactic of defense of an avowedly secular state against pressure for a yet greater role for religion. Perhaps the commission was already certain of what Pat Robertson would say.

The immense strength of a religious viewpoint is that it is capable of abolishing hard ethical problems if only we can correctly decipher the meaning of what has been revealed to us.[7] It is a question of having the correct "perspective." Philosophical "considerations" are quite another matter. The painful tensions and contradictions that seem to the secular moral philosopher to be unresolvable in principle, but that demand de facto resolution in public and private action, did not appear in the testimony of any of the theologians. While they disagreed with one another, they did not have to cope with internal contradictions in their own positions. That, of course, is a great attraction of the religious perspective. It is not only poetry that tempts us to a willing suspension of disbelief.

---

7. Once, impelled by a love of contradiction, I asked a friend learned in the Talmud whether meat from a cow into which a single pig gene had been genetically engineered would be kosher. His reply was that the problem would not arise for the laws of *kashruth* because to make any mixed animal was already a prohibited thing.

An Exchange

The exchange that follows was published in the March 5, 1998, issue of The New York Review.

HAROLD T. SHAPIRO, JAMES F. CHILDRESS, and THOMAS H. MURRAY, of the National Bioethics Advisory Commission, write:

We are writing to clarify what the National Bioethics Advisory Commission (NBAC) recommended in its report on *Cloning Human Beings*, which Richard Lewontin reviewed in *The New York Review*. Such reports are not easy to review, but no one could learn from Lewontin's review what the commission recommended or its reasons for doing so. Indeed, the review reveals much more about the reviewer's position on cloning and the relation of science, ethics, and religion than about NBAC's position or reasoning.

Lewontin reports correctly that the commission recommended a ban on cloning humans, but he fails to describe the nature, scope, and limits of that ban or to identify its rationale. Among its several recommendations, the commission recommended a temporary ban, through federal legislation, including a sunset clause, on the use of somatic cell nuclear transfer cloning to create children. Although the commission heard and considered many ethical arguments for and against human cloning, it based this recommendation solely on the ethical argument, in line with the available scientific evidence (which the review concedes), that the technique is not safe to use in humans to create children at this time. Our ethical concern around the issue of safety is a quite natural extension of the growing concern since Nuremberg for protecting human participants in scientific research and for avoiding the premature initiation of new clinical practices.

Protecting the children who might be born of somatic cell nuclear transfer cloning, and the women who might bear such children, was not, as we have already noted, the only moral argument the commission heard and considered. We heard concerns expressed that such cloning was a form of hubris or would lead inexorably to exploitation and even oppression; that it would result in children being treated as objects, even as commodities; that it would damage the integrity of families; that it would somehow threaten individuality and autonomy. While we were persuaded that some of these concerns deserved serious further reflection, this was not possible given the time constraints imposed upon us. As a result, the commission recommended a continuing national dialogue (to which the Lewontin review clearly contributes) that would focus on these and related ethical issues. Our hope was that such a dialogue involving "widespread and careful public deliberation" about a wide range of ethical and social concerns, together with new scientific evidence, would clarify society's ultimate view regarding the appropriate use of this new technology. Thus, the commission's recommendation regarding the *current* use of somatic cell nuclear transfer cloning techniques to create children rested only on the safety argument. The commission did not, contrary to Lewontin's interpretation, base its recommendation for a temporary ban on other concerns, such as objectification.

Lewontin also criticizes the commission for turning to religious scholars from some of the principal religious faiths in America to gain an understanding of their views on the matters before us and for including a chapter on "religious perspectives." We did so for two reasons. One was to access the resources they had developed over the years to deal with associated issues. Second, the simple fact that so many Americans look to these major religious faiths for moral guidance made it important for us to try to understand their perspectives.

America is a fascinating and complex nation of religious believers and nonbelievers of various stripes, with deep commitments to religious freedom and firm traditions regarding the noninterference of government in these matters. As a result, believers cannot look to the government to reflect back to them their particular religious beliefs, and no particular religion, therefore, can become the cultural project of the government. In our view, however, this does not bar the consideration in public debate of thoughtful arguments whatever their source. On the contrary, in a society such as ours, it is essential that we try to understand the thoughtful views of others. While recognizing that public policies in our society cannot be based on religious considerations alone, the commission wanted to learn from diverse theological and philosophical perspectives the positions taken and the arguments made about cloning humans. While religious traditions influence the moral views of many citizens, it is more relevant to note that moral arguments in these traditions often rest on premises accessible to citizens outside those traditions, and their norms and judgments often overlap with secular ones. Holding that "all voices should be welcome to the conversation" in our pluralistic society, the commission invited and discussed religious perspectives "in the spirit of sustaining a national dialogue," and in the belief that we all benefit from understanding the thoughtful views of others.

Finally, Lewontin distorts both theological and philosophical positions on cloning humans. He asserts that theologians attempt to "abolish hard ethical problems" and avoid "painful tensions." (We ignore his phrase "internal contradictions," which he includes along with "painful tensions," because both philosophers and theologians try to avoid "internal contradictions" in order to develop defensible positions.) We regret that Professor Lewontin was not present at our public meetings with religious thinkers, who were quite candid about the "painful tensions" they experienced in attempting to understand and explain what their traditions had to say about the difficult issues raised by these new developments. It was clear to us that many theologians as well as many philosophers, both in testimony to NBAC and in other contexts, do recognize these "painful tensions."

Neither group as a whole failed to appreciate the moral conflicts involved in cloning humans, in various scenarios, or in different public policies toward cloning humans, even if different thinkers resolved them differently. And NBAC's own reflections benefited greatly from both theological and philosophical perspectives and considerations.

RICHARD LEWONTIN replies:

The complaint of the members of the National Bioethics Advisory Commission, that I failed to report their final recommendation to the President, is a just one. They did, indeed, recommend a temporary ban on the creation of cloned children because it is not safe, and I should have said so explicitly. But that very recommendation, as well as the rest of their letter, simply underlines the shortcomings of the report to which I drew attention. As I pointed out at the beginning of my review, we do not ordinarily ask an *ethics* advisory commission for advice on the safety of medical research and procedures. Such matters are technical issues for the NIH or the FDA, and if there are serious doubts about safety, the ethical issue would seem to be settled. It is disingenuous of the commission members to suggest that these technical issues and their recommendation were somehow the central point of the whole affair, while they incidentally "heard concerns" about questions of individuality, exploitation, objectification, and the like and that they thought that "some of these concerns deserved serious further reflection." The commission asked for and received a great deal of testimony from theologians and ethical and political philosophers, and a large part of the 115-page report of the commission was taken up with these considerations. There were chapters devoted to "Ethical Considerations" and "Religious Perspectives," and the majority of the commission members were not even professionally competent to make judgments about technical biological issues. This was indeed an ethics inquiry in the broadest sense. The reliance of the commission on purely technical matters of safety for their recommendation seems a neat way of finessing the political problems raised by the ethical, but especially the religious, issues. After all, if it is unsafe, we really don't have to struggle over all the rest.

It is all the rest that raises unsettling social and political issues. A serious consideration of objectification could not avoid confronting the actual nature of relations between employer and employed and the degree to which most people have an actual choice about being objects whose value is calculated on the difference between their productivity and the cost of their wages. But that gets us into pretty deep political waters. A real concern with the false belief that genetic identity determines personal identity would involve an ethics commission in recommending some kind of serious state effort to enlighten people about the fallacy of biological determinism. And the commission certainly did not want to have to take a position on abortion, a position that would surely have led into problems of the relations between religion and the state. It was much the better part of valor to rest their case on safety and they have my genuine sympathy.

The discussion about religion in the letter from the commission members is rather contradictory. It is undoubtedly true, as they say, that religion informs either directly or indirectly the ethical and moral views of Americans. Indeed, I would even want to claim that Western secular moral philosophers cannot avoid the influence of biblical morality that has permeated the atmosphere of the culture in which they exist. But how can they, on the other hand, argue that "the fact that so many Americans look to these major religious faiths for moral guidance made it important for us to try to understand their perspective" while excluding testimony from precisely the major religious tradition, Protestant fundamentalism, that most deliberately enters public debate on issues of morality and state policy? The answer seems pretty obvious. It was a matter of safety again, this time political safety. The struggle over the role that religion is to play in forming state policy has never been more acute than it now is, and the commission needed to avoid getting embroiled in it.

Finally, I understand the state of moral philosophy differently from Shapiro, Childress, and Murray. Philosophers do indeed try to avoid contradictions, but these are what we may call "analytic" contradictions. That is, one is not allowed to say both "A" and "not-A" (unless the whole purpose is to demonstrate that such an analytic contradiction is contained in the very structure of the logic). The problem in moral philosophy, however, is not that there are analytic contradictions but that no system yet devised of constructing "ought" statements from basic axioms has been able to avoid practical contradictions like the conflicts of different but equal rights in theories of justice. That is, moral philosophy does not give us unambiguous directions about what to do in all situations. What religious revelation does is to provide the certainty that in all situations there is an unambiguously right thing to do, as given by Divine Law, and leaves only the question of how to know God's will. We may hear an inner voice or we may need help from an expert. We may be dissatisfied with what we hear and try to get a second opinion (the use of talmudic law is filled with second opinions). The text is there. The problem is to decipher it.

The contradiction facing the Bioethics Commission was that between practical politics and philosophical coherence.

Epilogue

Whatever the popular concerns about cloning may be, it is the commercial possibilities that drive the research. In the case of human cloning, money can be made from offering reproductive technology to individuals who for one reason or another may want to produce a clonal offspring. The trouble is that such a possibility immediately raises ethical, religious, and political questions that put the potential human cloner in a very unhappy position. When, not long after the announcement of the cloning of Dolly, an entrepreneurial physician, Richard Seed, proposed a program to make human clones on demand, he was greeted by immediate widespread opposition, including a ban on research aimed at cloning a human being in the United States. In the absence of any realistic technical basis for his promises, no investor seems ready to defy popular disapproval, so Dr. Seed's plans have come to nothing.

The realistic promise of money to be made from human cloning comes not from making an entire human being but from the possibility that human tissues or organs could be cultivated on a small-production scale, either for transplantation into sick or injured people or as a source for the factorylike production of specifically human proteins to be used in therapy. Tissue culture is an old technique, but its medical application has been prevented by several complications. Not all tissue types can be cultured and even those that can may not maintain their tissue specificity. Cells age and die because of changes that occur in their chromosomes and these must be prevented if tissues are to be kept in stock indefinitely. Moreover, culturing a single tissue is not the same as culturing an organ and it is often a whole organ, say a kidney, that is needed. Finally, tissue taken from one person will be rejected when transplanted into a second person unless the donor and recipient are immunologically quite similar, a similarity that depends on genetic similarity. It would then be extremely useful if cells from a known genetic source could be used to produce, on order, tissue of a specific type. It would be even more wonderful if such cells could be induced to differentiate into an entire organ.

The demand for such replacement tissues and organs is immense and would be the source of considerable cash for anyone who could satisfy it. It is the possibility of cultivating tissues, rather than of making babies, that has been the object of cloning research since Dolly. There have been four noteworthy advances within the last year. The first was the announcement in the summer of 1998 that fifty adult mice had been produced clonally from adult cells, rather than the egg cell used by Wilmut in his original sheep research. Thus, tissues that had already differentiated into a specific type could be returned to their uncommitted "totipotent" state and then develop into all possible organs and tissues. Then in the fall, stock of the Geron Corporation went from $6 to $23 a share the day they publicized the successful culturing of undifferentiated "stem" cells from a very early human embryonic stage. These cells, it was said, can be kept indefinitely in their undifferentiated state in petri dishes but then induced by appropriate stimuli to turn into specific tissues and organs. This was followed, in the spring of 1999, by a report of the successful culture of human stem cells taken from adults, followed by their differentiation into several different specialized tissues.

Most dramatically of all, it was announced at the end of 1998 that a successful embryonic culture had been made from cow cells in which the genes had been replaced by human DNA. There were cries of protest as visions of the horned Minos were conjured up, but presumably he had genes from both Europa and the Zeus bull. Since the culture did not progress beyond the early embryonic stage, nothing can be said about its humanity. The potential economic value of this last development is considerable, because it makes possible the production of large quantities of all kinds of human proteins that can be used in disease therapy, just as human insulin is now produced by bacteria carrying a single human gene.

Chapter 9

SURVIVAL OF THE NICEST?

"Survival of the Nicest?" was first published in The New York Review of Books *of October 22, 1998, as a review of* Unto Others: The Evolution and Psychology of Unselfish Behavior, *by Elliott Sober and David Sloan Wilson (Harvard University Press, 1998).*

IN *HIGHER SUPERSTITION*, a book remarkable both for its influence on the intellectual community and for its obtuse ignorance of the actual state of science, the authors told us that

> Science is, above all else, a reality-driven enterprise. . . . Reality is the overseer at one's shoulder, ready to rap one's knuckles or to spring the trap into which one has been led . . . by a too complacent reliance on mere surmise. . . . Reality is the unrelenting angel with whom scientists have agreed to wrestle.[1]

Any reader who wants to test this charmingly naive view of science should immerse himself or herself in the literature of evolutionary biology. Indeed, the immersion does not have to be very deep before the currents and countercurrents of ideology can be felt tugging

---

1. Paul R. Gross and Norman Levitt, *Higher Superstition: The Academic Left and Its Quarrels with Science* (The Johns Hopkins University Press, 1994), p. 234.

at one's understanding. *Unto Others*, a collaboration between Elliott Sober, one of the founders of the modern philosophy of biology, and David Sloan Wilson, one of the most creative theoreticians in evolutionary studies, wades into this turbulent stream at precisely the point where so many other adventurers have been swept away: the problem of the origin of altruistic behavior. Can natural selection have made us genuinely cooperative and unselfish in pursuit of the greater good of the many, or is apparent altruism nothing but an artfully disguised version of every man for himself? Have Professors Sober and Wilson really collaborated in order to spread enlightenment, or are they engaged only in a bit of academic career building, each using the other as a tool of their separate ambitions?

Darwinism, born in ideological struggle, has never escaped from an intimate reciprocal relationship with worldviews exported from and imported into the science. No one challenges the claim that evolutionary theory has had a wide effect on social theory. It is a cliché of cultural history that the explanation of evolution by natural selection served as an ideological justification for laissez-faire competitive capitalism and the colonial domination of the lesser breeds without the law. Nor are these evidences only of the quaint naiveté of the nineteenth century. Social Darwinism has had a continuous and vigorous life until today. Only three years ago a leading publisher of psychological monographs produced a book by a professor of psychology at a first-rank Canadian university claiming that the evident moral and cognitive superiority of Europeans over Africans was a consequence of natural selection in a cold rather than tropical climate. All the Africans got out of their experience of the survival of the fittest were greater libidos and longer penises.[2] The slightest suggestion, however, that evolutionary biology has imported some of its conclusions from social theory and political prejudice will be greeted by incredulity and indignation on the part of scientists convinced of the intellectual autonomy of the study of nature. Even those who insist that they concentrate on the "internal" history of science agree that Darwin's notion of the struggle for existence, and the consequent differential survival of those types with greater fitness for the struggle, owed a great deal to the economic and social theorists of the late eighteenth and early nineteenth centuries, such as Dugald Stewart and the Scottish economists; and they recognize that so-called "Social Darwinism" was a popular ideology long before the composition of *On the Origin of Species*.

Yet the mere mention in the pages of *The New York Review of Books* of this now conventional understanding was enough to invoke an irate response from one of the leading physical scientists of our time, Max Perutz, who assured readers that Darwin would have reached his conclusions irrespective of the intellectual atmosphere of the mid-nineteenth century because, after all,

2. Anyone who thinks this a caricature of a serious position should consult J. Philippe Rushton, *Race, Evolution and Behavior: A Life History Perspective* (Transaction, 1995).

those ideas were a revelation of natural reality. Scooping the authors of *Higher Superstition* by a dozen years, Perutz claimed that any suggestion of the importance of social context was evidence of the corrupting influence of Marxism on intellectual life.[3]

The pervasiveness of general worldviews in evolutionary biology, however, goes well beyond the relatively uncontroversial influence of nineteenth-century social attitudes on the historical origins of the science. They have informed at all times, including the present, the way in which evolutionary biologists describe the reality of nature. Hegel's lament in *The Lectures on the Philosophy of History* that "instead of writing history we are always striving to find out how we ought to write history" is more applicable to evolutionists than to German historians.[4] There is no simple and direct "truth" about how to understand the history of life on earth.

The descriptive facts of evolution are not at issue. Natural scientists agree that roughly three billion years ago life appeared from inanimate materials and that since then the organisms that have inhabited the earth have been connected to one another by a chain of ancestry that can be reconstructed from the fossil record and from molecular, physiological, morphological, and behavioral similarities among living organisms. When there are disagreements about facts they

---

3. See M. F. Perutz, "High on Science," *The New York Review*, August 16, 1990, pp. 12–15, and the exchange of letters between Perutz and me, *The New York Review*, December 6, 1990, pp. 69–70.

4. "*Statt Geschichte zu schreiben, bestreben wir uns immer zu suchen, wie Geschichte geschrieben werden müsse.*"

are a consequence of the necessary ambiguities of historical reconstructions from limited observations.

There are brief, bloody struggles over fine details of ancestry, such as, for example, whether human beings have a more recent common ancestry with chimpanzees or with gorillas, or whether one's favorite fossil primate is in the direct line of human ancestry. There are, as well, disagreements about whether or not the appearance of new species is often the consequence of very short but dramatic periods of splitting and morphological change, separated by long periods of boring stasis (the theory of "punctuated equilibrium") and whether the disappearance of old species is often the result of literally world-shaking events like a collision with a meteor. Some ideological predispositions may enter in these latter cases, pitting those who are committed to the view that history is a long and gradual incremental movement against those who see revolution as a vital element in historical change.

In the end, however, such disagreements do not create deep and long-lasting divisions among evolutionists. Sometimes, as in the case of the human-chimpanzee-gorilla puzzle, it is agreed that no observations can ever resolve the question, and, anyway, what difference does it really make? Sometimes, as for the arguments about punctuated change and catastrophic extinction, a pluralistic agreement is reached with only local skirmishes persisting about whether this or that particular sequence of fossils fits more closely the gradualist or catastrophist scenario. Individual careers are made by emphasizing one sort of case rather than the other, but evolutionary biology as a whole is not threatened with reformulation.

It is when we move from evolution as narrative to evolution as universal history, from Hegel's category of Original History to his Reflective History, that the predispositions of ideology come to dominate the science. The practitioners of evolutionary biology, no less than their philosophical and literary hangers-on, seem determined to impose on the history of life a single unifying principle or viewpoint that is said to be what evolution is "really all about." And not only evolution as a process but every detail of the life activities of every species that has ever lived, certainly including *Homo sapiens*. There are three such unifying themes that plague evolutionary biology today, and all three converge in the problem of altruism that is the main subject of *Unto Others*.

The first claimed universal in evolution is that it is an optimizing process. External nature poses problems for organisms, problems of life maintenance and of reproduction. Different individual organisms "solve" these "problems" differently and those who solve them "best" leave more offspring who also inherit their method of solution through their genes. Organisms propose and nature disposes. As a consequence the best solution becomes more common and finally comes to characterize the species. Thus, natural selection, a process of conjecture and refutation, leads ineluctably to the optimum solution to the problem. The claim of optimization arises in part from an unreflective literal reading of a nineteenth-century slogan, "survival of the fittest," but also from a widespread misunderstanding of a technical issue. Given a particular environment, each genetic type in a population has some probability of survival and reproduction, what we call the *fitness* of the type. These fitnesses can be averaged, weighting each by how frequent the genetic type is in the population, to produce a number called the *mean fitness* of the population. Evolution is a change in the frequencies of the different types from generation to generation, with the result that the mean fitness of the population changes from generation to generation. It turns out that if natural selection is the only force operating on the population, the frequencies change in such a way that the mean fitness of the population only increases and never decreases.

The standard metaphor is therefore evolution as a mountain climber. Mean fitness is like the altitude in a mountain range, and natural selection is like an inner compulsion of a climber to climb yet higher and higher. Hence evolution by natural selection is seen as a maximizing or optimizing process driving species to greater and greater fitness. But that conclusion is a misunderstanding of the metaphor. A mountain range, including the Fitness Mountains, contains many peaks of different heights, and a climber who wishes to ascend the highest peak must be able to see into the distance and choose an appropriate path to Everest. Unfortunately, the mechanism of natural selection does not allow such global behavior and forces the population upward along the particular local slope on which it finds itself.

In this case the mountain climber cannot see beyond the end of his alpenstock and has no way of "knowing" that higher peaks exist elsewhere, that there are even better possible solutions than the present one.

The false view of natural selection as a process of global optimizing has been applied literally by engineers who, taken in by a mistaken metaphor, have attempted to find globally optimal solutions to design problems by writing computer programs that model evolution by natural selection. In fact, none of these schemes of "genetic algorithms" has yet succeeded in solving a design problem that was not already solved by more conventional methods. The most talked-about example is the famous "Traveling Salesman's Problem," which has implications for any design question that involves connecting a large number of points by straight pathways. This problem calls for a generalized Willy Loman to visit a large number of cities that are spread out over the map in an irregular way, taking the shortest (or cheapest) possible total path, without visiting any city twice. Vast amounts of computing are required to get even an approximate solution, and genetic algorithms are worse than other techniques.

Believers in evolution as an optimizing process assert more than that natural selection is a rule for finding the best. They claim, too, that all properties of all species are a consequence of direct natural selection for those characteristics. None is accidental, none is a failure of optimum adaptation, none is the epiphenomenal consequence of other, perhaps optimal, features, none is the ineluctable outcome of being flesh. The mountain climber never tumbles down a crevasse, dizzy from a deficiency of oxygen or an excess of brandy. It has been seriously proposed by a respectable evolutionary ecologist, for example, that the holes that rot in the trunks of older trees are really a favorable adaptation because small animals are attracted to live in them and these animals help to spread the seed of the tree. A more extreme form of this hypothesis is that even the fact of death is a consequence of natural selection, designed to prevent those of us who are reproductively worn out from eating the bread of our still-fertile offspring.

Evolutionists have not always been such Panglosses, however. The Whiggish belief in evolutionary progress by survival of the fittest that characterized the exuberant expanding capitalism of the last half of the nineteenth century gave way, after the unimaginably bloody slaughter of the First World War and the economic desperation of postwar depressions, to a rather less optimistic view of the march of history. With essentially the same facts of natural history at their disposal as are now available, the evolutionists of the first half of the present century differed from present-day evolutionists in their characterization of the process. *Evolution: The Modern Synthesis*, edited by Julian Huxley in 1942, which brought together the leading evolutionists of the time, was filled with the consciousness of historical contingency. While some argued that the differences between species were a direct consequence of natural selection, others argued that reasonably often "the race is not to the swift, nor the battle to the strong, nor riches to the wise man, but time and chance happeneth to all." H. J. Muller, winner of a Nobel Prize for his demonstration of the artificial induction of mutations by ionizing radiation, thought that deleterious mutations would accumulate more and more over time in populations as a consequence of purely random variations in individual fecundity, with a consequent steady degeneration in the species.[5]

During the last thirty years, however, despite the fact that the technical literature of evolutionary genetics has emphasized more and more the random and historically contingent nature of genetic change over time, the literature of natural history, of ecology, and of behavioral evolution, and the growing body of popularizations produced by evolutionists, philosophers, and science writers have again become unrelentingly optimalist. This change has not occurred because a mass of new facts has forced us to a new vision of reality. On the contrary, the development of very sophisticated statistical methods has been required to detect any signal of natural selection above the din of random DNA variation that has been observed by modern molecular evolutionists. Where does the faith in optimality come from? Certainly not from inside science.

The second article of ideological orthodoxy, virtually unchallenged at present by any student of evolution, is

---

5. An undoubted phenomenon known to evolutionary geneticists as "Muller's Ratchet."

that the individual organism is the object seen directly by natural selection. That is, any argument that some characteristic has been favored by natural selection must be of the form that individual members of a population who display the characteristic will leave more offspring than those who have other traits. What is being explicitly denied is that characteristics favorable to the population as a whole will evolve by natural selection, except as a secondary consequence of the greater fitness of individuals over others within the population. So, for example, we are not allowed to claim that linguistic communication between humans was favored by natural selection by arguing that a group of protohumans who could talk to each other would be at an advantage in warfare or hunting over other groups who were without language. Somehow it would have to be argued that a single individual with a greater linguistic capacity than others would, as a consequence, leave more offspring.

Moreover, as a fortune I once received in a Chinese restaurant pointed out to me, "the best talkers don't necessarily make the best listeners," so we need two sorts of selection simultaneously, which only compounds the problem. If we were to allow, however, that a characteristic might spread through a species because local groups that accidentally possessed it somehow took over the species, the problem would be solved. This possibility of group selection has been regarded as anathema by nearly all evolutionary biologists, although entirely without empirical evidence. The obvious hypothesis is that the exclusive concentration on the individual as the unit of selection is a direct transferral onto evolutionary theory of the central role of the individual as actor in modern social and economic thought. If there is a benefit to the group, it is simply a manifestation of the invisible hand.

The difficulty posed by a combination of panselectionist optimizing theory with a commitment to the individual organism as the sole locus of natural selection becomes obvious when we consider altruism. The evolutionary theorist means by altruism a particularly strong form of benefit to others in which that benefit is at the expense of some harm to the altruist. Organisms do indeed seem to sacrifice themselves for others' benefit. Mothers and fathers sacrifice food and rest for the benefit of their children, and siblings can usually count on each other for support in time of need. But altruism extends outside the family. Soldiers throw themselves on grenades to save their comrades, and even in New York a man may give up a taxi in favor of a pregnant woman.

The problem of altruism is regarded by some evolutionists, in particular those who identify themselves as sociobiologists, as the outstanding problem of evolutionary biology and its solution as the outstanding contribution of sociobiological thought. Hillel is said to have once been challenged by a prospective convert to provide the essence of his religion while standing on one leg, to which the sage replied, "Do not unto others what you would not have them do unto you. All else is commentary." For sociobiologists, too, unselfishness is the key and all else is commentary. Their real program is not simply to explain social phenomena as a product of evolution, but to demonstrate the *universality* of optimizing natural selection as the explanation of all features of all organisms. The ambition of evolutionists to possess a universal rather than a contingent truth is in danger of being thwarted by the seeming irrationality of unselfishness. Ought natural selection not have expunged such counterreproductive behavior? Doesn't natural selection favor the selfish, those who maximize their own life and reproduction, at the expense of others? Where have we gone astray?

The answer offered by the most trendy evolutionary theory is that we have not gone down far enough in our search for reality. The group is not the unit of selection, nor even the individual organism. In the third article of ideological orthodoxy, it is the *gene* that is the real object and beneficiary of natural selection. Organisms are merely the temporary and mortal vehicles of the immortal DNA, instruments of the selfish genes. In Richard Dawkins's disturbing metaphor, we are "lumbering robots" under the direction of the genes that "created us, body and mind."[6] If we take the gene's-eye view, then what appears as altruism at the level of the individual can be understood really as a form of selfish manipulation by the devious genes. If it is the "purpose" of a gene to maximize its own reproduction, then this can be accomplished in different ways. One is the usual pathway of influencing the development and

6. See Richard Dawkins, *The Selfish Gene* (Oxford University Press, 1976).

physiology of the individual organism, causing it to out-reproduce other organisms.

Another way, called *kin selection*, is to make me help the reproduction of other organisms that carry genes of the same kind, even at the expense of my own reproduction. Because my close relatives are likely to carry genes identical to mine, acquired from our common ancestor, then a gene can do just as well by increasing the welfare of several of my kin at my expense. For example, from the gene's point of view, two of my brothers or sisters are worth one of me, because on the average they carry about half of the same genes as I do. So, if a sneaky gene can induce me to give up some resources to them without utterly bankrupting myself, its own purposes would be well served.

Yet a third way, producing altruism to unrelated strangers, is to lend fitness at interest. I will do a favor to you, and therefore to your genes, even at some cost to me, hoping that your genes will make you reciprocate at a future time, to the benefit of both batches of DNA. An example offered by sociobiologists is the saving of a drowning man, at some risk to the lifesaver, who hopes for future reciprocation. The trouble with this example is that the last person I would count on to save me from drowning is someone who could not swim well enough to save himself. Nor do the books quite balance in the case of the soldier who throws himself on a grenade, but nothing is perfect and the supposition is that on the average over all situations things will work out.

To make it work, however, we need some appropriate mechanism of biological mediation between genes and behavior. Can there be genes "for" being nicer to your brothers and sisters than to your second cousin once removed, or for casting your bread upon waters whose ebb is balanced by their flow? Can genes really modulate the structure of the central nervous system to produce just the right contingent behavior? Nothing is known by way of an answer to this question, and, more important, there is no program of empirical work on the central nervous system intended to make these formal speculations into concrete anatomy and physiology. At this point in the wrestling match the "unrelenting angel of reality" seems to have its wings pinned to the mat.

At first sight, *Unto Others* appears to be a reformulation of the now orthodox view of the evolution of altruism. It is, however, a great deal more subversive than that, for, if its alternative scheme is taken seriously, evolutionary biologists should stop characterizing the process as one in which genes drive organisms to develop particular characteristics that maximize their fitness. Genes cease to be the "real" beneficiaries of natural selection; individual organisms cease to be the sole "real" objects directly seen by the selective process; there is nothing that is optimized or maximized (except in a tautological bookkeeping sense), and the entire process of evolving a particular behavior does not require that there be genes "for" the behavior.

It might be supposed that some external ideological commitment might be invoked to explain Sober and Wilson's radical revision of the theory of the evolution of altruism. Surely Kropotkin's replacement of competition by cooperation in his own theory of evolution was the consequence of the self-conscious application of a principled general position. The case of *Unto Others* is more easily explained, however, as arising from institutional and idiosyncratic causes. Intellectual work is supposed to be a combination of originality and hard thinking. Unfortunately, there is some contradiction between these, at least in evolutionary theory. Careers are often made either from an ambitious but poorly thought out originality, or a skillful but mechanical analysis of a well-worn theme. *Unto Others* is precisely that combination of radical reexamination of a system of explanation, an examination from the roots, with a rigorous technical analysis of both biological and epistemological questions that we all are supposed to engage in. What marks off their intellectual production is not its ideology but the seriousness with which they have taken the intellectual project.

The hinge of Sober and Wilson's argument is a rejection of the prejudice that natural selection must operate directly solely on individuals. They point out that groups of organisms may also be the units of differential reproduction and they provide examples from natural history that can only be understood if, in addition to the survivorship of individuals, the survivorship of entire collections of organisms is taken into account. In fact, the effect of selection within groups may be the opposite of the effect between groups, with the group effect dominating the entire evolutionary process.

The story of the rabbit in Australia is a clear example. Rabbits were introduced into Australia in the nineteenth century in order to allow colonial country squires to continue the English tradition of huntin' and shootin'. Unfortunately the squires' marksmanship could not keep up with the reproductive rate of the rabbits, who quickly became a serious pest and competitor with sheep. An attempt was made to control the rabbit population by introducing an infectious lethal disease, myxomatosis, which spread rapidly and, at first, decimated the rabbit population. But then the rabbits began to increase again. As was to be expected, rabbits resistant to the disease organism had appeared, presumably by random mutations, and genetic resistance spread through the species by natural selection. In addition, however, the disease organisms had also evolved to be less virulent so that even susceptible rabbits were less damaged by the newly evolved pathogens.

This seems odd, because more virulent strains grow more quickly and should eliminate less virulent strains when competing within rabbits. The trouble with being virulent, however, is that, having taken over a rabbit, the virus kills it. Myxomatosis is spread from rabbit to rabbit by a mosquito and mosquitoes do not bite dead rabbits. So, having taken over the subpopulation of viruses by natural selection within a rabbit, virulent viruses have guaranteed their own eventual elimination by failure of that subpopulation to be transferred to new hosts. Ethnic cleansing has been the pathway to national suicide and the more benign eventually survive globally.

In considering groups as units of selection, it is important not to take too impoverished a view of what constitutes a group. It need not be a spatially defined unit, as in the myxomatosis case, where all the virus particles contained physically within one rabbit form a selected group that then may contribute to new groups in new rabbits. Groups may be delimited by any shared property, such as food preference, temporal pattern of activity, gender, social class, or e-mail list. Anything that sorts individual organisms will do, including kinship. Thus Sober and Wilson swallow up kin selection as a special case of their more general theory. For group selection to operate, all that is required is that there be collections of individuals that interact with each other in some way separate from the interactions of other individuals in other groups, and that subsequently the groups contribute differentially to the next generation.[7]

In fact, it is hard to imagine any real biological species that is not broken up into such "trait groups" according to many different traits, so that a mixture of individual-level selection within collections and group-level selection between them ought to be the rule. The exploitation of plants as a food resource for insects is an example. Female moths make a choice of plants on which to lay eggs, depending on various physical and chemical properties of the plants. There is variation in preference among females within a moth species, partly

---

7. Even e-mail groups enlarge or diminish over time as a consequence of the attractiveness of their subject matter and the stimulus offered outsiders by the content of their interchanges. This occurs even though the least thoughtful and interesting participants often produce the longest and most frequent messages.

as a consequence of genetic differences between moths, but also as a result of exposure to different plant chemicals during their own development as caterpillars. All the caterpillars that hatch out on a given plant are a "trait group," having in common the consumption of a particular plant with particular characteristics. During their lives as caterpillars some will be more efficient or voracious than others, and so survive better to adulthood. Their genetic type will therefore increase within a group.

But this fact alone is not sufficient to predict whether their type will increase in the species as a whole. That increase or decrease depends in part on the commonness or rarity of the plant variety on which they survived and on how choosy they will be as adults when they are ready to lay eggs. A type that survives extremely well in competition on a rare plant will decrease in the species as a whole if its members insist on laying eggs on that rare plant, but may increase if they are more catholic in their tastes for plants. The ultimate pattern of plant consumption for the moth species as a whole is, then, a consequence of a mixture of individual and group selection in which the balance depends in part on how widely the individuals born on one sort of plant will disperse to other sorts of plants to produce the next generation.

Using the trait-group approach, Sober and Wilson show quite convincingly that species may evolve altruistic behavior provided that the frequency of altruistic types within groups has an effect on the contribution of the group as a whole to the next generation of the species. If some group has, by chance, a higher frequency of altruistic individuals, and if the consequence is a larger number of offspring for the group as a whole, then even though there is some selection against the altruists within each group, altruism may come to characterize the species. Demographic studies of Native American groups have confirmed that, as might be expected, so-called "war chiefs" who were elected to lead their fellows into battle had smaller numbers of children as a consequence of their greater likelihood of death in combat. Success in war presumably enhanced group survival, however, so the altruistic act of the war chief, sacrificing himself for the group, would nevertheless lead to a survival and spread of the altruistic institution.

The modern ideology of population control predisposes us to think that small families are good for the individual, good for the family, and good for the species as a whole. Large families are harder to support and demand a sharing out of limited domestic resources, and we all know that the world is being ruined by overpopulation. But such an argument confuses the present with the past, owners with peasants, and well-being with numbers. For most of the history of agricultural Europe, landowning families were well advised to have the largest possible families in order to maximize, by marriage connections, the land and military force on which they could depend. Even today, in rural agricultural India, family prosperity is increased, not decreased, by extra children whose unpaid agricultural labor often means the difference between a net profit or loss to the family.[8] Moreover, the question of the increase or decrease of numbers in a population should not be confused with their prosperity. The prosperous few have more than once succumbed to the ill-fed masses.

Obdurate genic selectionists will respond that, irrespective of the mechanical details, all that matters in the end is which genes increase and which decrease. Groups, like individuals, are here today and gone tomorrow. Only the gene remains, so we need to consider only the differential reproduction of different genes, however that may be mediated. By definition, a gene is more fit in evolution if it leaves more copies in the next generation. But this bookkeeping trick confuses causes and effects, or, rather, eliminates material causes by reifying statistical effects. First, it confuses random changes with selective changes. In all countries at all times surnames become more or less common, and even extinct, because of variation in reproductive rates among families that are purely at random with respect to the names themselves. Martin is the most common French surname, whereas Bonaparte is practically nonexistent. This is a consequence of the fact that when surnames were created Martin was a common given name, but also that Martins have left a lot of children and Bonapartes only a few.

---

8. For a detailed economic analysis of why family-planning programs are rejected by Indian small farmers, see Mahmoud Mamdani, *The Myth of Population Control* (Monthly Review Press, 1973).

We would not want to say, however, that these names in themselves had some causal efficacy in reproduction, whatever the social and biological properties of their carriers may have been. That is, we would not want to ascribe Darwinian fitness to a name. There are many causes for an increase in genes that have nothing to do with the direct physiological consequences of bearing those DNA sequences. Different families leave different numbers of offspring for reasons that are at random with respect to a particular gene, so in a finite population of organisms there will be random changes in the frequency of genes from generation to generation. These random changes eventually cause the complete loss of one form of a gene and its chance replacement by another form.

This process of "neutral evolution" characterizes most changes in DNA during evolution, and its very selective neutrality is the basis for the reconstruction of the past relationships of living species. Sometimes unselected genes are swept rapidly to a high frequency because they happen to be on the same chromosome as a gene that is itself the object of natural selection. To assign fitnesses to genes that increase or decrease by chance or because they hitchhike on the chromosomes of other genes is a tautology that completely obfuscates the actual causal events.

Second, there is a confusion of the level at which causal action is occurring. The effect, at one level, of processes occurring at a different level may give an incorrect picture of what is happening. A famous example was the suspicion during the 1970s that the graduate school at the University of California at Berkeley was discriminating against women, because the overall acceptance rate of male applicants was clearly higher than for females. When each department was looked at separately, however, men and women were being accepted at the same rate at which they applied. The apparent discrepancy arose because women were disproportionately applying to the departments with the lowest overall acceptance rates. That is, they were engaged in riskier behavior than men, so they failed more often when averaged over all groups, although not within any given group.[9]

A gene with no deleterious effect on its carrier, but which is present in a species that lives only in the rich soil on the sides of slumbering volcanoes, will not decrease in frequency within that species (and may even increase), but its overall prospects are dim. To claim that selection is operating against this gene just because the species within which it occurs lives in a risky environment is an example of what Sober and

---

9. Sober and Wilson's numerical example is as follows. Suppose that 90 women and 10 men apply to a department with only a 30 percent acceptance rate (Philosophy?). In an unbiased process 27 women and 3 men succeed. In a second department 10 women and 90 men apply but this department accepts 60 percent of its applicants (English?) without bias so 6 women and 54 men succeed. On the average over the whole collection, 100 men and 100 women have applied, they have been accepted without bias, yet only 33 women as compared to 57 men were accepted. Indeed the effect could be seen even if there were a bias *in favor* of women within departments. Suppose each department accepted 2 more women and 2 fewer men. There would then still be only 37 women and 53 men accepted. This is an example of what is known as "Simpson's Paradox," arising from the fact that the probability of an average is not the same as the average of the probabilities.

Wilson call the "Averaging Fallacy." To call it a "fallacy," however, misses the point. The issue is not an analytic one as the word "fallacy" implies, but a metaphysical one about causal reality. If one continues to insist that the gene is what "really" matters, or professes a complete lack of interest in material mediation in favor of computing outcomes, then there is no fallacy. Once again the angel of reality seems to have abandoned us.

A large part of *Unto Others* is taken up with a classic problem in philosophy and psychology that is analogous to the evolutionary question of whether the appearance of altruism at the individual level is really selfishness at the genic level. Is human altruism really egoism, or even pure hedonism, in disguise? Egoism is entirely self-directed. The egoist asks only "Is it good for me?" The answer, of course, may involve whether it is good for others, who may also be egoists, because it is beyond dispute that we may sometimes benefit ourselves by means of benefiting others. The egoist must spend a certain amount of time doing cost-benefit analyses, but with confidence that some bread is worth casting on some waters.

The important point of the claim for egoism is that the welfare of others enters the calculation only instrumentally. Hedonism, on the other hand, is a particular psychologistic and somewhat unreliable variety of egoism in which the actor asks only "Does it feel good?" without making a calculation. If what I was told about cod-liver oil when I was a child is true, then the senses are an unreliable guide to objective benefit, and assuredly villains smile and smile even outside of Denmark, so how we feel about situations may fool us. Still, risky as hedonism may be, our apparent altruism may be only doing what makes us feel warm inside. That is, we always act only to achieve gratification, but we have been brainwashed by our upbringing and social circumstances into being gratified through being nice. Others then reap the benefit of our ultimate self-ishness. A refusal to be tempted by the devil may, after all, only be giving in to the ultimate ego satisfaction of achieving sainthood.

Sober and Wilson are too well versed in the history of this ancient problem to claim they can know the truth of the matter. They know they cannot rule out either egoism or hedonism as the "real" source of human altruism, and they recognize these theories as being ideological predispositions that are not capable of unambiguous empirical tests. They are clearly tempted by the psychological experiments that are most easily interpreted as showing the existence of real empathy for others' circumstances. For example, a group of test subjects is told that a group of their fellow students need help in going over their schoolwork. Some of these needy students are described in such a way as to make it easier for the subjects to empathize with them while others are described in a more alienat-ing way. In addition, half the subjects are offered a mood-enhancing experience (such as pleasant music) whether or not they helped a student, while the other half were not offered such a reward. In the absence of the reward more subjects helped those with whom they were empathic. The other subjects, despite the promise of an unconditional mood-enhancing experience, still went to the trouble of helping those with whom they felt more empathy. So the claim of the experimenters is that real empathy counts. But Sober and Wilson admit that this sort of experiment is rather unconvincing, and they can offer no suggestion for a better one. They confess that their temptation to believe in a real effect of empathy, in addition to any hedonistic motivation, arises out of their own a priori predisposition to believe in irreducible altruism.

In the end, Sober and Wilson are entirely forthright in saying that they have consciously adopted a pluralistic perspective. In their view evolution occurs at many levels of causation, from the gene to the population. There is natural selection of genes, of individuals, and of whole groups, and all of these are going on in the evolution of altruism. In psychology they accept the existence of egoistic, hedonistic, and irreducibly altruistic motivations for apparently altruistic behavior. To the extent that they support attempts to explain phenomena at lower levels of causation, at the genic rather than the organismal, or the organismal rather than the population level, they do so only for strategic reasons. They are methodological reductionists, because to ascribe actions at higher levels without an attempt to explain them at lower levels invites an indiscriminate obscurantist holism that is the enemy of understanding. I share their pluralism, but, of course, that may really be because it makes me feel good.

Chapter 10

GENES IN THE FOOD!

"Genes in the Food!" was first published in The New York Review of Books *of June 21, 2001, as a review of* Genetically Modified Pest-Protected Plants: Science and Regulation, *a report by the Committee on Genetically Modified Pest-Protected Plants, Board on Agriculture and Natural Resources, National Research Council (National Academy Press, 2000);* The Ecological Risks of Engineered Crops, *by Jane Rissler and Margaret Mellon (MIT Press, 1996);* Stolen Harvest: The Hijacking of the Global Food Supply, *by Vandana Shiva (South End Press, 2000); and* Pandora's Picnic Basket: The Potential and Hazards of Genetically Modified Foods, *by Alan McHughen (Oxford University Press, 2000).*

1.

IF THE NINETEEN recent books and fifteen-pound stack of articles that confront me as I write are any measure, then nothing is more productive of food for thought than thoughts about the production of food. The introduction of methods of genetic engineering into agriculture has caused a public reaction in Europe and North America that is unequaled in the history of technology. Not even the disasters at Three Mile Island and Chernobyl were sufficient to produce such heavy and effective political pressure to prohibit or further regulate a technology, despite the evident fact that uncontained radioactivity has caused the sickness and death of very large numbers of people, while the dangers of genetically engineered food remain hypothetical.

It is out of the question to review this vast literature in its entirety, so I have chosen four recent characteristic examples from the pile. One is a report and set of recommendations from the font of American scientific legitimacy, the National Academy of Sciences/National Research Council. A second, *The Ecological Risks of Engineered Crops*, is a partisan but temperate case for

344

the dangers of genetic engineering in agriculture, produced under the auspices of a long-established political action group, the Union of Concerned Scientists. The third, *Stolen Harvest*, is an unremitting indictment of genetic engineering, on moral, cultural, and economic grounds, especially as it applies to the third world. The fourth, *Pandora's Picnic Basket*, is the only example I could find of the opposite prejudice. It is a defense of genetic engineering in agriculture and a bitter attack on the apparatus of government regulation written by an agricultural scientist, an inventor of transgenic varieties whose life was made difficult by government regulation.

Whatever fears I might have of possible allergic reactions to food produced from genetically modified organisms, they are not more unsettling than the allergies induced in me by the quality of the arguments about them. What are we to make of a major issue of science and public policy in which a physicist bases her opposition to genetic engineering on "the recognition in the *Isho Upanishad* that the universe is the creation of the Supreme Power meant for the benefits of (all) creation"[1]; or a professor of agricultural economics who, in the course of trying to convince us that technology is good for farmers, conveniently makes the elementary error of confusing total household income of farm families with income *from* farming[2]; or a senior research

---

1. Vandana Shiva, *Stolen Harvest: The Hijacking of the Global Food Supply*, p. 17.

2. NRC report, *Genetically Modified Pest-Protected Plants: Science and Regulation*, Appendix A, p. 220.

scientist working in plant breeding at a major public university who ridicules the need for regulatory oversight of new kinds of foods by citing the introduction of macaroni and cheese on a stick that was announced in his local newspaper[3]? And these examples are, alas, characteristic of what has been written. Even the most judicious and seemingly dispassionate examinations of the scientific questions turn out, in the end, to be manifestoes. We are presented with a paradigm of Julien Benda's *trahison des clercs*; but *The Treason of the Intellectuals* was concerned with the corrupting effects of ideological passions on intellectuals. Ideological passions about potatoes? It gives one to think.

The uproar about so-called genetically modified organisms (GMOs) has been the direct consequence of the development of a radically new way to manipulate heredity. Human beings have been genetically modifying organisms since the first domestication of plants and animals. The results of those ancient modifications have been organisms that are not only very different from their wild ancestors, but are in many characteristics the very opposite of the organisms from which they were derived. The compact ear of maize with large kernels adhering tightly to the cob is very useful in a grain that needs to be gathered and to be stored for long periods, but a plant with such a seed head would soon disappear in nature because it could not disperse its seed. The

---

3. Alan McHughen, *Pandora's Picnic Basket: The Potential and Hazards of Genetically Modified Foods*, p. 198.

history of domestication is precisely the history of the genetic modification of organisms to make them most "unnatural."

Until recently the method for producing new varieties of plants or animals has been to search for desirable variants and to propagate them selectively. The naturally occurring variation within species can also be augmented by matings with closely related species that do not ordinarily interbreed in nature, but will do so under conditions of domestication. So classical methods of plant and animal breeding have included "unnatural" transgression of species boundaries. But the use of the genetic variation available only from closely related organisms limits what can be accomplished precisely because they are closely related and therefore quite similar. Moreover, introducing genetic variation by crossing between organisms is imprecise. A cross between two varieties is indiscriminate in the hereditary characteristics that are transmitted. Thus if one attempts to introduce disease resistance into an especially high-yielding variety of wheat by crossing that variety with one that has the disease resistance but not the high yield, the result will be a variety with improved resistance but lower yield. The ideal of the plant or animal engineer is to be able to remake the heredity of an organism to order, so as to produce just those variants that the occasion seems to require.

Apparently the secret of genetic engineering was known to the ancients. Genesis 30 tells us that in order to retain the services of his son-in-law Jacob, who was apparently quite good at animal husbandry, Laban agreed to let him keep all the speckled and streaked goats and sheep that were born in the flocks that he tended. Jacob, the ur-biotechnologist, then peeled some twigs to make them speckled and streaked and held them up before the eyes of the plain-colored ewes just as they were about to conceive. This produced the desired result and Jacob became very rich indeed.

Being of little faith, we seem to have lost the twig trick, but have invented a new one. Modern genetic engineering consists in extracting the DNA corresponding to a particular gene from a donor organism and then inserting it into the cells of a recipient in such a way that it becomes incorporated into the recipient's genome. This insertion can be carried out by coating tiny metal particles with the DNA and shooting them into the recipient cells or by first putting the DNA into microorganisms and then infecting the recipient with them. If the source of the DNA is a distant species that cannot be intercrossed with the recipient, the engineered result is said to be a transgenic organism. The donor and recipient need not be anything like each other for the trick to work.

Thus the human gene for insulin has been successfully inserted into the genome of bacteria, and these bacteria, grown in industrial vats, are now churning out human insulin for the market. Despite the fears about the human ingestion of the products of genetic engineering, there has been no widespread concern about the large numbers of diabetics who are injecting bacterially produced insulin twice a day, even though a number of people have reported more or less severe difficulties in adjusting their blood sugar levels using this product of genetic and chemical engineering.

The chief use of transgenic DNA transfers in agriculture up to the present has been to provide crop plants with resistance to insect pests or to make the plants resistant to herbicides used to control weeds. The resistance to insects has been created by inserting into plants the genes coding for powerful toxins, the Bt proteins, from a bacterium, *Bacillus thuringensis*. When insects begin to nibble the plants, they ingest the Bt toxin and die. Resistance to herbicides has also been transferred into a variety of crop plants from bacteria, as well as from a variety of unrelated plants that happen to be resistant to particular chemicals. One of the ironies of the current struggle over GMOs is that advocates of organic farming practices who strongly oppose the introduction of transgenic crops containing the Bt genes have for many years promoted the dusting of the bacteria themselves on plants as an organic substitute for chemical insecticides.

While an irony, it is hardly the contradiction that proponents of GMOs suggest. The dusting of a toxin on the outside of plants, from which it could be washed away, is not the same thing as having the plants manufacture it internally. Although pest and herbicide resistance have been the main focus of transgenic engineering until now, anything seems possible. What makes the technique so attractive and so productive of anxiety is that any gene in any species can be transferred to any other species. Of course, some of these transfers will be harmful or even lethal to the recipient organism so that no practical use can be made of them; but there are no general rules to tell us what will work.

The critical point is that there is no limit to what could be done if it were worth someone's while to do it. Hundreds of plant varieties created by genetic engineering have been tested under guidelines approved by federal agencies, and several dozen transgenic varieties are commercially available, including corn, cotton, squash, potatoes, canola, soybeans, and sugar beets. It has only been six years since the first transgenic crops were planted commercially, yet now more than 20 percent of maize acreage in the United States is planted in transgenic corn and worldwide there are about 100 million acres sown in a variety of transgenic crops, including cotton and soybeans.

The usual reaction of the federal government to widespread public agitation about public health and environmental issues is to tinker with already existing regulatory procedures. When scientific questions are involved, federal agencies or Congress will often request that the National Academy of Sciences, through its research arm, the National Research Council, produce an expert report to guide regulatory policy. Sometimes, however, the Academy will act even without such a request. The National Academy of Sciences is a self-perpetuating body of the American scientific elite that provides technical advice to the government. Its leadership, conscious of its legitimacy as a font of supposedly disinterested and expert opinion on scientific questions, will sometimes arrange for National Research

Council reports unbidden, on the assumption that their weight of authority will have an effect on public policy.

The NRC has issued, without a formal request, several reports on genetic engineering since 1974, when it became clear that recombinant DNA techniques would be important as tools of genetic research and technology. Three of those reports have been directly concerned with the application of the techniques in agriculture, one in 1987 on the release of GMOs into the environment, one in 1989 on the safe field testing of transgenic varieties, and, in 2000, *Genetically Modified Pest-Protected Plants*, which includes a discussion of both the environmental issues and threats to human health.

The creation of a scientific report on a contentious issue presents a special difficulty. On the one hand the drafting committee must include representatives of various constituencies with opposing views. So the committee that wrote the new report included academics involved in genetics, economics, and agriculture, a representative of a public interest environmental action group, a lawyer who helps clients to obtain regulatory approvals, and a state government environmental regulator. On the other hand, there cannot be a majority and a minority report, since after all we are dealing with Objective Science, and scientists either know the truth or they don't. NRC reports always speak with one voice. Such reports, then, can produce only a slight rocking of the extremely well gyrostabilized ship of state, no matter how high the winds and waves. Any member of the crew who mutinies is put off at the first port of call.

While usually artfully concealed, the machinery of

forced consensus is apparent in the pest-protected plant report. The economist on the committee, Erik Lichtenberg, clearly felt that the sorts of regulation recommended by the report were not worthwhile and, indeed, would have costs not justified by any claimed benefits. He and his cost-benefit analysis are quarantined in an appendix and referred to only in a footnote: "This appendix was authored by an individual committee member and is not part of the committee's consensus report. The committee as a whole may not necessarily agree with all of the contents of appendix A." Of course, appendix A is merely economics, while the "committee as a whole" must "necessarily agree with the contents" of the rest of the report or it wouldn't be a scientific report. In fact, the committee could have discounted the appendix on substantive grounds. Like so much of cost-benefit analysis, it fails to take account of the fact that the costs, possible ill-health, fall on different parties than the benefits, profits to corporate entities who produce the inputs into agriculture. More fundamentally, it avoids the deep problem that to provide a quantitative balancing of the books, the costs and benefits would have to be assessed in the same currency, while it has never been possible to come to a general agreement on the dollar cost of sickness and death.

2.

There are five general issues that are in contention in the struggle over GMOs. Three of these, threats to

human health, possible disruption of natural environments, and threats to agricultural production from a
more rapid evolution of resistant pests, comprise the
agenda of the NRC report. The other two, disruption
of third-world agricultural economies and principled
objections to "unnatural" interventions, are deliberately
excluded. Page 2 of the report states in italics: "*The
study does not address philosophical and social issues
surrounding the use of genetic engineering in agriculture, food labeling, or international trade in genetically
modified plants.*" In analyzing the risks of GMOs the
committee follows a general principle established in
previous Academy reports, a principle that it regards
as fundamental, namely that it is the product and not
the process that matters. For the NRC it is irrelevant
whether a variety has been produced by conventional
genetic manipulations or by transgenic transfer of DNA.
What counts is whether the new property of the result-
ant organism is harmful to health or the environment.

The NRC authors point out, quite properly, that the
conventional methods of breeding, including sexual
crosses between species that do not ordinarily cross in
nature, might produce varieties with some heightened
toxicity to humans or other species, or with unusual
invasive abilities, or with greater resistance to pests that
would hasten the evolution of more effective pest species.
Jane Rissler and Margaret Mellon, in their extremely
informative *The Ecological Risks of Engineered Crops*,
give many examples of new troublesome weeds that
have arisen from the hybridization of crop plants with
their wild relatives and several where rare wild species

have been driven to extinction by hybridization with crop plants.

Indeed, the only examples we have so far of the adverse effects of agricultural varieties on any animal or plant species in nature, including on human health, have been from conventionally bred organisms or from the introduction of invasive species from distant geographical areas, or from foods like peanuts or milk to which some people are naturally allergic. So if the usual products of agricultural practice already provide numerous examples of adverse effects, why is there the massive popular and political anxiety centered on genetically engineered crops in particular? None of the authors of the reports and books seems to have noticed that if it were really only the product and not the process that matters, then nothing has changed. The NRC report itself provides a protocol for protecting consumers against new food toxins and allergens (i.e., substances causing allergies) that applies irrespective of the genetic method used in variety development and which makes use of the already existing federal apparatus for the approval of new plant varieties.

First, one asks whether a new substance is found in parts of a plant that consumers eat or with which workers come in contact. If not, the substance is "exempt from health concerns." If it is found in such parts, then does it have chemical properties common to many allergens? If it does, then safety assessment is needed. If not, then is it similar to other substances that people eat? If not, then again we need safety assessment. The real problem revealed in the NRC report, although it did not seem to bother the panel, is that the data on which "safety assessment" is currently based are not produced by the federal agencies themselves but are provided by the very parties who are asking for approval to distribute the new variety in the first place. Moreover, no one seems to have noticed that there is, in fact, an aspect of the process of genetic engineering that does make that process unusually likely to produce unpredictable results.

All the attention has been paid to the physiological effect of the gene that has been put into the recipient, but none to the effect of where it is inserted in the recipient's genome. Genes consist of two functionally different adjacent stretches of DNA. One, the so-called structural gene, has information on the chemical composition of the protein that the cell will manufacture when it reads the gene. The other, the so-called regulatory element, is part of a complex signaling system that concerns where and when and how much protein will be produced. When DNA is inserted into the genome of a recipient by engineering methods it may pop into the recipient's DNA anywhere, including in the middle of some other gene's regulatory element. The result will be a gene whose reading is no longer under normal control.

One consequence might be that the gene is never read at all, in which case it will probably be bad for the recipient and will never be part of a useful agricultural variety. But another possibility is that the cell will now produce vast amounts of a protein that ordinarily is produced in very low amount, and this high concentration could be toxic or be involved in the biochemical production of a toxin. Yet another possibility is that a toxic substance that used to be produced only in one part of a plant, not ordinarily eaten, could now be manufactured in another part. Tomatoes are delicious, but you would be ill-advised to eat the leaves and stems because they contain toxins. It is not impossible that a genetically engineered tomato might, by bad luck, start to produce these toxins in the fruit. Thus the process of genetic engineering itself has a unique ability to produce deleterious effects and, contrary to the recommendations of the NRC report, this justifies the view that all varieties produced by recombinant DNA technology need to be specially scrutinized and tested for such effects. Exactly how one would go about doing that, in view of the unknown nature of the danger, is uncertain. Even extensive testing on a variety of animals provides no guarantee of safety since there are plant substances that are toxic to some species and not to others.

As yet no one that we know of has been poisoned by a transgenic plant. There have been a couple of close calls, however. The most widely cited case is the Brazil nut protein produced by a transgenic soybean. In some subsistence agricultural communities, for example in West Africa, diets are severely deficient in an essential amino acid, methionine. Brazil nuts produce a protein that is rich in methionine and so it was thought that inserting the appropriate gene from Brazil nuts into soybeans would provide an easy fix for West African malnutrition. Unfortunately the Brazil nut protein is known to be allergenic and the transgenic soybean proved to be so as well, so the variety was never released.

Proponents of recombinant DNA technology like Alan McHughen point to this case as a proof that self-policing by a variety developer can be counted on to avoid disaster. One's confidence in self-policing is somewhat diminished, however, by the realization that the allergenic properties of the protein were well known before the Pioneer Hi-bred seed company ever started to develop the variety in the first place. At some point they must have realized that the Food and Drug Administration would have refused approval of the variety even under our present system of regulation. How one wishes for a transcript of the discussions in the company board room.

A major part of the NRC report and the entirety of Rissler and Mellon's book are concerned with ecological issues in the broad sense. One anxiety is that "superweeds" will be produced, dominant plants that will spread en masse either through cultivated fields or through natural habitats. Sometimes what is meant by "weeds" is unwanted species that are growing in cultivated fields. At other times these are confused with introduced invasive species like the European purple loosestrife that has taken over so many American wetlands. There are no known examples of hybrids between cultivated plants and wild relatives becoming superweeds that have destroyed natural habitats, largely because too many of the characteristics selected during domestication make cultivars—cultivated varieties of plant species—dependent on the tender loving care given to them by farmers. Nor will the addition of a gene conferring herbicide resistance or pest resistance change that dependence. Plants growing in natural habitats are not subject to herbicides, nor are they attacked regularly by the hordes of predatory insects attracted to the concentrated free lunch offered in cultivated fields.

On the other hand, more difficult weeds of cultivated fields certainly will evolve if herbicide resistance becomes incorporated by natural crossing into species that are already weeds. The fear of superweeds is promoted by the metaphor of "escape" used to describe the passage of an engineered gene into a wild species. The image is of the mad scientist (or not-so-mad germ warfare biologist) who has created a virulent disease organism, ready at any moment to create a major epidemic unless it is rigorously confined to the laboratory or, better yet, destroyed. But transgenes are not spread like microbes, entering the body from outside. They are transmitted by reproduction of the entire genome of an organism, and if a cross occurs between an engineered plant and a wild relative, the result is an offspring that is hybrid in every respect, including all those characteristics that make cultivated varieties so ill-adapted to survival in nature, such as their demands for unnaturally high levels of nitrogen fertilizer.

The opponents of GMOs are not alone in the misuse of the image of "escape." McHughen, in his manifesto against the regulation of biotechnology, claims that spatial isolation of fields in which transgenic crops are growing is utterly useless because the transgenes have already escaped onto roadsides and other fields through seed that is inevitably spilled from sacks, trucks, and machinery in the very process of transportation and planting. But this small amount of spilled seed is irrelevant. What is properly of concern is not the escape of a virulent infection, but that a constant rain of millions of pollen grains produced by hundreds of acres of a transgenic crop will over and over produce hybrids with weedy species at the margins of cultivated fields and eventually result in a now weedy form that will be unusually invasive or competitive.

3.

Most of what is written about GMOs is quite parochial, concentrating on their effects in North America and Europe. While we expect nothing more from the National Research Council or from an indignant Canadian plant engineer, the general lack of interest in the effects of biotechnology on the third world seems in contradiction to the rather moralistic tone of the public discourse. Predictably, the most famous example of a piece of biotechnology that is supposed to be good for subsistence agricultures is cited by McHughen, but, unfortunately, it does not do the work intended. A serious problem of nutrition in some rice-producing regions, causing blindness, is a lack of vitamin A. A transgenic variety, Golden Rice, has been created with the promise that if it is ever cultivated, it will provide the missing vitamin. But Golden Rice—not to be confused with Green Revolution rice—does not, in fact, provide vitamin A. It is enriched in beta carotene, a precursor of the vitamin (hence the golden color), which can only be converted to vitamin A in the body of an already well-nourished person. The developers of Golden Rice have not dealt with this problem in their publicity releases. Rissler and Mellon have a brief final chapter entitled "International Implications," but these are largely the extension of the ecological risk arguments already made for the United States and do not deal with promised nutritional benefits.

The only recent book that deals with the effect of agricultural biotechnology in the third world and embeds it in a more general discussion of agricultural technology in general is Vandana Shiva's *Stolen Harvest*.[4] Shiva is what is called a "cult figure" for opponents of GMOs, but her book will give a detached observer more the impression of a cheerleader. She might have used her knowledge of Indian agriculture and her immense prestige among environmentalists to provide a credible up-to-date analysis of the effects of agricultural technology and market structures on third-world economies. Instead, she has produced a conjunction of religious morality, undeveloped assertions about the cultural

---

4. The classic work on the effects of biotechnology in the third world is Calestous Juma's *The Gene Hunters* (Princeton University Press, 1989), which remains the basic source for an economic and historical analysis of the effect of agricultural technology in Africa and Asia. Because the work is a dozen years old, it antedates most of the actual development of GMOs and the immense growth of public discourse and anxiety about the subject.

implications of Indian farming, unexplained claims about the nature of the farm economy in India and how biotechnology destroys it, and unanalyzed or distorted scientific findings. *Stolen Harvest* is an opportunity squandered.

So with no further elucidation we are told that seeds and biodiversity are "gifts from nature and their ancestors" that Indian farmers have received; that "food security is not just having access to adequate food. It is also having access to culturally appropriate food"; that "the smoke from the mustard oil used to light the *deepavali* lamp acts as an environmental purifier." While Shiva makes the undoubtedly correct claim that conversion to high-yield Green Revolution varieties has resulted in less fodder for cattle and less green manure for fields and has displaced the culture of legumes, other vegetables, and fruits, she nowhere explains why Indian farmers have engaged in this self-destructive activity and how the global structuring of agricultural trade in combination with the internal economy of India has driven them to it. Indeed, she never shows that Indian farmers are worse off than before the introduction of agricultural technology.

Most disheartening of all, Shiva's reports of facts are not always as complete as they need to be. In a discussion of genetically engineered soybeans she writes that "infants fed with soy-based formula are daily ingesting a dose of estrogens equivalent to that of 8 to 12 contraceptive pills." It turns out, however, that the soybeans contain a nonsteroidal estrogen whose physiological activity is less than one thousandth the activity of the standard hormone. I learned this fact, not mentioned by Shiva, by consulting the very article from which she says her dosage figure was calculated.

The real present danger to third-world agriculture from transgenics is elsewhere. Much of the agricultural economy of these countries depends on growing specialty commodities like lauric acid oils used in soaps and detergents, once found only in tropical species. Now, with recombinant DNA, these are produced by canola. Why buy palm oils from the politically unstable Philippines, where 30 percent of the population depends on it economically, when we can grow it in Saskatchewan? Caffeine genes have been put into soybeans. Why not Nescafé from Minnesota?

No unequivocal conclusions can be drawn about the overall effect of genetic engineering technologies. It is clear that any manipulation of organisms, whether by conventional means or by genetic engineering, poses some danger to human health, to present systems of agricultural production, and to natural environments. All of these potential effects have led to a fairly effective apparatus of government regulation whose chief deficiency is its dependence on data supplied to it by parties whose prime concern is not the public good but private interest. Nothing is significantly changed in this situation by the introduction of genetic engineering. The technology provides a method for transferring a specific gene into a crop, rather than the uncontrolled mixture of entire genomes that takes place when two varieties or species are crossed. On the other hand the

362 random disruptions of regulatory genes of the recipient that may take place are totally uncontrolled. On balance, it is impossible to say whether we have achieved greater or lesser control over the unintended consequences of mucking around with nature.

We find ourselves in a puzzling situation. None of the books on the subject of GMOs gives us any reason to think that the known dangers to human health and natural ecosystems posed by agriculture have become radically greater because of the introduction of genetic engineering as a technique. Nor do we even have a single case of a catastrophe that might have engendered widespread public anxieties.

Yet in North America, and much more so in Europe, there is a widespread, passionate, and politically effective opposition to the use of recombinant DNA techniques in agriculture. Only a rare defensive newspaper advertisement paid for by the Council for Biotechnology Information speaks against the general consciousness, and we all know whom they represent. Is this just another chapter in MacKay's *Extraordinary Popular Delusions and the Madness of Crowds*? A hint at the answer can be found in a series of full-page newspaper advertisements created by the Turning Point Project, a coalition of over sixty political action organizations including Food First, the Sierra Club, and Greenpeace. One set of advertisements had as headlines:

Unlabeled, untested . . . and you're eating it

Biotechnology = Hunger

Genetic Roulette

Who plays God in the 21st century?

Just the usual anti–genetic engineering stuff? Consider another set:

Can industrial agriculture feed the world?

The myth of efficiency

America's last family farms?

Well, it's not just genetic engineering that is being opposed. It's really part of the organic food ideology. The next set of headlines makes a new connection:

Global Monoculture

Globalization *vs.* Nature

Invisible Government

Somehow we have moved from DNA to the WTO, but we are not finished. The progression is completed with

Monocultures of the mind

If computers in schools are the answer, are we asking the right question?

The Internet and the Illusion of Empowerment

E-Commerce and the Demise of Community

Techno-Utopianism

Now we understand the Turning Point Project. They're a bunch of Luddites. Right century, but wrong movement. The followers of the unseen King Ludd and Captain Swing from 1811 to 1830 were industrial and rural laborers thrown out of work or trying to live on poverty wages, who destroyed knitting and threshing machines that had displaced their labor. Their objection to technology was not ideological but pragmatic. If we want to find the nineteenth-century equivalent of the sources of Turning Point consciousness, we must find it in the movement that began with Blake and ended with Rossetti, Ruskin, and the pre-Raphaelites, in the call to arms against the dark Satanic Mills:

> *Bring me my bow of burning gold!*
> *Bring me my arrows of desire!*
> *Bring me my spear! O clouds,*
> *   unfold!*
> *Bring me my chariot of fire!*
>
> *I will not cease from mental fight,*
> *Nor shall my sword sleep in my hand,*
> *Till we have built Jerusalem*
> *In England's green and pleasant land.*

That nineteenth-century discontent was the reaction of a middle class repelled by the spiritual and physical ugliness created by a surging industrial capitalism to which they sensed no attachment. One might think that because the rise of industrial capitalism occurred so long ago and the culture it created has become so much the basis of European and American life, any truly popular new romantic movement against it would be inconceivable. But what was then a struggle against the rise of its dominance is now a struggle against its last consolidation in spheres of life that seemed set apart.

Until twenty years ago there were four intimate aspects of our personal lives that we assumed to be produced by individual artisanal activity. They were medicine, entertainment, sport, and vegetables. Some penetration of capital into those spheres had, of course, occurred but they were invisible to us. Since then our family physician has become a corporate health care practitioner; television, popular music, books, and film are owned by a few major conglomerates; baseball players are paid millions by owners who are paid millions by television networks who are paid millions by advertisers; and now Monsanto wants to tell me what to eat.

One consequence has been the creation of a false nostalgia for an idyllic life never experienced. I once bought a new computer in a large computer supermarket in a shopping mall in Boston. The salesman offered to carry the machine out to my car in the parking lot and as we approached the rear of the car, he spotted my green and white Vermont license plate. "Vermont!" he said, "That's where I really want to live." "Oh," I replied, "have you spent much time in Vermont?" "Oh, no," he said, "I have never been there." The independent family farmer, tilling the soil, in touch with nature, making decisions about what and when to plant and harvest from his craft knowledge, sitting down at dinner to a groaning board of home-grown victuals prepared by his aproned wife, is our last connection with an authentic life. We want to preserve it. Unfortunately, we are a hundred years too late and GMOs are the wrong target. To understand the situation we need more mental fight and fewer arrows of desire.

The history of American and European agriculture over the last hundred years has been a history of the increasing dominance of industrial capital over farming. In 1900 the inputs into farming were predominantly self-produced. The farmer saved seed from the previous year's crop to plant, the plow and tillage machinery were pulled by mules fed on forage grown on the farm, 40 percent of planted acreage was in feed crops, and livestock produced manure to go back on the fields. Now the seed is purchased from Pioneer Hi-bred, the mules from John Deere, the feed from Exxon, and the manure from Terra. The rise in purchased industrially produced inputs has had two effects. A major increase in yields per acre has driven down the price paid to farmers for their product. Simultaneously the farmers' costs of production have risen. There has been no escape from this dilemma for an individual farmer. Because the price paid for a farm product is determined by the aggregate production from all farms, no individual farmer can push prices up by holding down production. Thus he must increase production when other farmers do, but the result of all these individually economically rational acts is mass suicide. Smaller and smaller margins between farm income and expenses have led to increasing farm debt and bankruptcies.

The consequence of the growing dominance of industrial capital in agriculture for the classical "family farm" has been the progressive conversion of the independent farmer into an industrial employee. More and more farm operators and their spouses are only part-time farmers, trying to support their farming from outside income. That is why the confusion between farm family income and income from farming in the appendix to the NRC report is so misleading. In 1997, 60 percent of farm operators were also employed off the farm and 40 percent worked at alternative employment for more than two hundred days a year. They work as truck drivers, salespeople, secretaries, and factory workers. Car companies now put their assembly plants in the rural counties of the farm belt to take advantage of this labor force. It is not Jerusalem that has been built in the green and pleasant land, it is the dark Satanic Mills.

The creation and adoption of genetically modified organisms are the latest steps in this long historical development of capital-intensive industrial agriculture. Roundup Ready herbicide-resistant soybeans have been created by Monsanto so that farmers will be able to use its powerful herbicide, Roundup, while at the same time buying Monsanto seed. The farmers accept the cost of the new variety and its chemical partner because the use of such a powerful general weed killer will reduce the number of herbicide treatments or mechanical tillage passages through the fields, freeing them for the hours in the automobile assembly plant that they need to keep their farms. For the farmer there is no escape from engineering, whether it be mechanical, chemical, electrical, or genetic.

The text type, Sabon, was designed by the son of a letter-painter, Jan Tschichold (1902–1974), who was jointly commissioned in 1960 by Monotype, Linotype, and Stempel to create a typeface that would produce consistent results when produced by hand-setting, or with either the Monotype or Linotype machines.

The German book designer and typographer is known for producing a wide range of designs. Tschichold's early work, considered to have revolutionized modern typography, was influenced by the avant-garde Bauhaus and characterized by bold asymmetrical sans serif faces. With his Sabon design, Tschichold demonstrates his later return to more formal and traditional typography. Sabon is based upon the roman Garamond face of Konrad Berner, who married the widow of printer Jacques Sabon. The italic Sabon is modeled after the work of Garamond contemporary, Robert Granjon.

In Sabon, Tschichold's appreciation of classical letters melds with the practicality of consistency and readability into a sophisticated and adaptable typeface.

Sabon is a registered trademark of
Linotype-Hell AG and/or its subsidiaries

Printed and bound by R. R. Donnelley & Sons,
Harrisonburg, Virginia

Interior design by Red Canoe, Deer Lodge, Tennessee
Caroline Kavanagh
Deb Koch